THE FUTURE OF
LOW-BIRTHRATE POPULATIONS

Everywhere in the world populations of largely European origin are currently experiencing not only their lowest ever fertility levels, but what seems likely to be their longest ever periods of fertility at below replacement levels. Although it is widely assumed that the fertility of these countries will return to replacement levels within 30 to 35 years there is at present no empirical evidence that this will happen.

The inevitable demographic results of this fertility pattern are an older age structure and a decline in numbers. Many see this as leading to labor shortages and wage inflation, even to weakened national defense and the disappearance of European peoples and culture.

Numerical declines and older age structures are inevitable in today's low-birthrate populations. But they are unlikely to be either as great or as disruptive as commonly anticipated. Moreover, the policies proposed to avoid such demographic developments are clearly unsuitable. The inevitability of these changes – new in human history – must be accepted before societies can adjust to them and realize the benefits that inhere in them.

The Future of Low-Birthrate Populations assesses the demographic situation, the policy alternatives and the significance of future changes in fertility and mortality rates, and then discusses what can be done to minimize the losses and maximize the gains attendant upon a dwindling and aging population.

Prior to his recent retirement **Lincoln H. Day** was Senior Fellow in Demography at the Australian National University.

D0218277

THE FUTURE OF
LOW-BIRTHRATE
POPULATIONS

Lincoln H. Day

London and New York

First published 1992
Paperback edition first published 1995
by Routledge
11 New Fetter Lane, London EC4P 4EE

Simultaneously published in the USA and Canada
by Routledge
29 West 35th Street, New York, NY 10001

Typeset by J&L Composition Ltd, Filey, North Yorkshire
Printed and bound in Great Britain by
Biddles Ltd, Guildford and King's Lynn

British Library Cataloguing in Publication Data
A catalogue record for this book is available from the British Library.

Library of Congress Cataloging in Publication Data
Day, Lincoln H.
The future of low-birthrate populations / Lincoln H. Day.
p.   cm.
Includes bibliographical references and index.
1. Fertility, Human—Europe.   2. Demographic transition—Europe.
3. Europe—Population policy.   I. Title.
HB991.D38   1992   91–46079
304.6′2—dc20   CIP
ISBN 0–415–08034–7
0–415–12704–1 (pbk)

To Alice
wife, collaborator, friend

CONTENTS

vii

CONTENTS

FIGURES

ix

TABLES

FOREWORD

Nearly a third of a century ago, I wrote an article (L. H. Day 1960) warning of the threat to the environment and the quality of human life posed by continued population increase in the U.S.A., and, by implication, in industrialized countries generally. My wife and I later developed this into a book (L. H. and A. T. Day 1964) in which we sought to do three things: (1) show in some detail how continued population increase lowers the quality of life, (2) rebut the arguments in favor of continued population growth (namely: (a) that there are no limits, science-will-save-us; (b) that population increase is necessary to the maintenance of economic wellbeing; (c) that more people are needed for military defense; and (d) the eugenic argument that, if a cessation of population growth is, indeed, necessary, then let all those 'inferior' people undertake what is necessary to its achievement, while we 'superior' people go on having as many children as we want), and (3) propose practical means to the cessation of population increase in the U.S.A. The final sentence of that book gives something of the flavor of our argument:

> In short, what we must have – and what our recommendations have been aimed at – is a society in which: (1) no unwanted child is born; (2) the decision to bear or not to bear a child is made solely by the potential parents: and (3), most important of all for the goal of a stable population, this decision is made in a social and cultural context in which a family of three children is considered large.

Now that fertility in the populations of industrialized countries around the world has begun to follow the course we advocated, all those years ago, it is timely to look into what might be the consequences. This I have tried to do here. I have my own values

xiii

and priorities (which the reader should have no difficulty in sorting out); but I have tried to look as dispassionately as possible at the situation – guided by my training and research experience in demography and sociology, as well as by a fair bit of reading and thinking over the years concerning ecological issues, city planning, economics, and social gerontology. In all of this, but particularly in social gerontology, which is her current area of research and professional specialization, I have gained much from conversations with my wife. I see difficulties and challenges; but none insurmountable or such as would justify the hand-wringing about older age structures and numerical declines that one so frequently encounters these days.

Although inevitable, the numerical declines and older age structures expected for today's low-birthrate populations are unlikely to match current expectations as to either their magnitude or their seriousness. Moreover, the policies proposed to avoid population decline and aging are either irrelevant, ineffectual, or manifestly undesirable. Prudence would seem to argue, therefore, that we accept the inevitability of these changes (which are new in human history) and set about doing what is required both to adjust to them and to realize the benefits that inhere in them.

In the preparation of this book, I have used not only the resources of my home institution, the Australian National University, but also those of three fine facilities in Washington, DC: the Library of Congress, the library of the United States Public Health Service, and the impressively comprehensive and up-to-date specialized library of the American Association of Retired Persons. In addition, I was helped in the early stages of the work by an appointment as Visiting Scholar at the Population Reference Bureau, Washington, DC; and in the later stages by an appointment as Scholar-in-Residence at the Rockefeller Foundation's idyllic Bellagio Study and Conference Center, Bellagio, Italy. To each of these institutions and their able staffs I express my gratitude.

I am also grateful to David Rabin of the Georgetown University Medical School and Thesia Garner of the United States Bureau of Labor Statistics for sharing ideas with me and providing me with data.

My intellectual debts are many: to my teachers and former students, and to such members of the unusually stimulating intellectual community I have found in Canberra as: Bryan and Anne Furnass, Christabel Young, Terence and Valerie Hull, Don

Rowland, Don and Joan Anderson, John Barnes, Allan Martin, Mark Diesendorf, John Harris, and the late Roger Bartell.

I owe a particular debt to my wife, Alice Taylor Day. At this point in a 'foreword,' authors often thank their wives for assistance with typing. But Alice's contribution was more cerebral than that. Her knowledge, insight, and wisdom were of very special assistance throughout. The typing (and writing) I did myself.

REFERENCES

Day, L. H. 1960. The American fertility cult. *Columbia University Forum* 3(3), 4–9.

Day, L. H. & A. T. Day 1964. *Too many Americans*. Boston, Mass.: Houghton Mifflin (paperback edition, New York: Dell, 1965).

PREFACE

Everywhere in the world, populations of largely European origin are currently experiencing not only their lowest-ever fertility levels, but what seem likely to be also their longest-ever periods of fertility at below-replacement levels. This circumstance has come about in the absence of depression, war, or any other likely contributing factor. Although the projections currently in use by both the U.N. (1986) and the World Bank (Zachariah & Vu 1988) incorporate the assumption that the fertilities of these countries will return to replacement levels by about the end of the first quarter of the 21st century, there is, at present, no empirical evidence that they will.

Some of this low fertility may be the result of nothing more than a postponement of births. But it is by no means implausible to suppose that below-replacement level fertility could be, as one observer has put it, the 'logical, perhaps largely unavoidable, end stage of full demographic transition ... that individuals given full control over their fertility will find better things to do with their time than fully replace themselves' (Zinsmeister 1986, 590). Whether individuals' evaluations of 'better' will match social evaluations or accommodate to society's needs is, of course, a matter of some debate (see, e.g., the papers in Davis, Bernstam & Ricardo-Campbell (eds.) 1986).

The inevitable demographic results of this fertility pattern are an older age structure and a decline in numbers. But while European age structures have been getting older for a long time, numerical declines – at least as a result of low fertility – are of more recent origin. Nevertheless, numerical declines have already been occurring in Denmark, Germany, and Hungary (Population Reference Bureau 1987), and they can be expected to commence within the next decade in Austria, Belgium, Luxembourg, and Sweden

(Zachariah & Vu 1988). Barring marked increases in fertility or the commencement and indefinite persistence of positive net migration, declines in population numbers will eventually occur in every country in Europe (excepting Albania) as well as in those countries outside Europe whose populations are of mostly European origin.

Not surprisingly, these demographic changes have occasioned some alarm (summarized in Teitelbaum & Winter 1985); less in professional than in lay circles, to be sure, but to some extent in both. The prospect of numerical declines in combination with still older age structures generates a host of fears – of, for example, labor shortages and wage inflation, weakened national defenses, shortages of intellect, declines in national 'vigor,' race suicide and the disappearance of European culture. It is a heady brew.

Because they can be expressed numerically, demographic phenomena (and, for the same reason, economic) are often accorded a causal significance beyond their due. In discussion of these matters, the causal nexus between the low fertilities of economically developed European populations and the deleterious social consequences these fertility levels will presumably give rise to tends to be, at best, simplistic; often, it is not even specified. Yet, surely, any claim as to causal association between demographic conditions and such phenomena should incorporate into the equation all manner of nondemographic factors – factors, in many instances, of arguably greater relevance than the merely demographic. Demographic conditions do set limits, but in matters like these, such limits are necessarily broad.

For most of the fears expressed about European fertility decline the outlook is at the least uncertain. This is because both aging on the scale expected and the shrinking of national population size (especially as a result of low fertility) are so new to human experience. Mortality will have some effect, but the extent of the demographic changes will depend in the main on how far fertility falls, and how long it remains there. The social consequences of these demographic changes – of older age structures and declining numbers of people – will depend, however, not so much on demographic dimensions as on a variety of essentially nondemographic factors: on, for example, the types of social policies in force; the distribution of wealth and income; the availability of health services, housing, and public transportation; and on such conditions of the physical and social environment as air quality, noise, personal safety, and opportunities for both social interaction and the enjoyment of privacy.

What I propose to do in this book is assess the demographic situation and the policy alternatives for dealing with it, then consider the likelihood of significant future changes in fertility and mortality levels, and, in conclusion, assess the likely losses and possible gains to these populations attendant upon their having both older age structures and smaller numbers.

REFERENCES

Davis, K., M. Bernstam & R. Ricardo-Campbell (eds.) 1986. *Below-replacement fertility in industrial societies*, supplement to Vol. 12, *Population and Development Review*.

Population Reference Bureau 1987. *World population data sheet*. Washington, DC.

Teitelbaum, M. S. & J. M. Winter 1985. *The fear of population decline*. San Diego, Calif.: Academic Press.

U.N. (United Nations) 1986. *World population prospects: estimates and projections as assessed in 1984*. ST/ESA/Ser.A/98, New York.

Zachariah, K. C. & M. T. Vu 1988. *World population projections, 1987-1988 edition*. Baltimore, Md: Johns Hopkins University Press.

Zinsmeister, K. 1986. Review in *Population and Development Review* 12(3), 589-92.

1

THE DEMOGRAPHIC SITUATION

THE EUROPEAN SHARE OF THE TOTAL

Because no population in history has yet experienced such an alteration of its demographic structure, discussion of the consequences of the demographic course of events posited in the current projections for low-birthrate populations is necessarily speculative. There is, however, one fear that the prospect of this change has given rise to about which there can be no question: the European share of total world numbers will inevitably decline. The process is already under way – less a result of low European fertility, however, than of the continuation, elsewhere, of much higher fertility and the higher growth rates that emanate from it. Low European fertility only hastens this process; it does not account for it.

Europeans were some 18 per cent of the world total in 1750. By the middle of the 20th century, following upon two centuries of unprecedented growth, they had increased their share to some 29 per cent (estimated from data in Durand 1967, Table 1, assuming half of the Russian population to have been European in both periods, and Europeans, in the later period, to have constituted 90 per cent of the combined populations of the United States and Canada, 50 per cent of the populations of Middle and South America, and 77 per cent of the population of Oceania). During that period, world population more than trebled and Europeans emigrated in sizeable numbers to every continent. Between 1950 and 1985, European numbers (on the same assumptions as to their proportions within regions) increased another 50 per cent (from some 722 to some 1091 million), but their share of the world total dropped to 23 per cent (U.N. 1986, Table 4). If the fertility assumptions of the United Nations' 'medium variant' projection are

1

realized, European numbers (using the same assumptions as to proportionate shares by continent) will have increased another 32 per cent by 2025, but their proportion of the world total will have declined further still – to 18 per cent (*ibid*., Table 4); to, that is, about what it was on the eve of the modern period of rapid population increase.

It is doubtful whether such a change in relative proportions could, itself, have much effect on European lifestyles and living conditions. Nor would it be likely to cause or quicken the pace of a decline in European influence or the demise of European civilization. Whatever their share of the world's total, a billion or so Europeans cannot help but figure prominently on the world stage – not to mention the fact that it took far fewer than that number (and at a much lower level of technology and standard of living) to create and develop European civilization in the first place. The more probable demographic threat to European lifestyles and civilization may well be not too few Europeans, but too many.

RECENT TRENDS IN THE FERTILITY OF LOW-FERTILITY POPULATIONS

In the countries under consideration here, the lowest fertility levels preceding those of the present period were reached in the Great Depression of the 1930s. On the basis of the 'net reproduction rate' (a measure of the extent to which, under a given set of age-specific birth rates, the average woman would reproduce herself during her period of childbearing), the mid-1930s population of Germany was failing to reproduce itself by 30 per cent a generation, that of Britain by 23 per cent, of France by 13 per cent (calculated from date in U.N. 1954, Table 21), and of Austria, which had the lowest fertility of all, by a whopping 33 per cent (calculated from data in Carr-Saunders 1936, insert between pp. 122 and 123) (Chart 1.1).

But because of their relatively youthful age structures and their subsequent increases in fertility, in none of these countries, except France, did a decline in total numbers actually occur – and France's modest decline (280,000 – less than seven-tenths of a per cent of the population (U.N. 1951, Table 3)) occurred only because that country's exceptionally long (at least 150-year) history of the individual practice of birth control had produced an age distribution that, from the standpoint of demographic stability, was relatively undistorted.

2

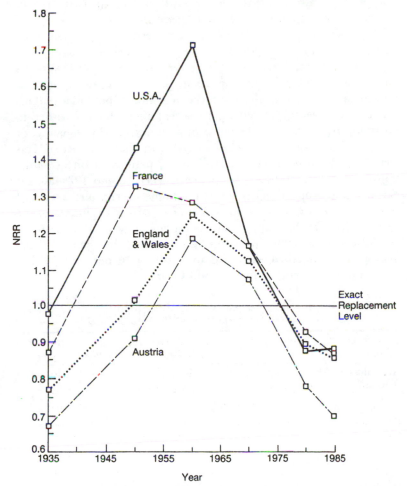

Chart 1.1 Changes in net reproduction rates, 1935–85: selected low-fertility countries

Fertility rose markedly in these countries in the period following World War II, and then again commenced to decline, about the beginning of the 1960s. In the years since, it has never returned to either the unusually high levels of the post-World War II period or the more usual, but nonetheless higher, levels of the pre-Depression period. Moreover, this time around, the pattern of low and declining fertility cuts a wider swathe, for it includes those

European countries, mostly in the East and South, that did not experience the declines of the earlier period.

The differences between the fertility patterns of today and those of the 1930s in these countries are substantial and significant. Today's levels are lower, they have remained at these low levels longer, and, perhaps most important, they appear to represent a fundamental change in the numbers of children potential parents desire (or, at least, are willing to accept in preference either to obtaining an abortion in the event of an unintended pregnancy, or to exercising greater contraceptive effort in the first place). This change shows up particularly well in the pattern of births at the upper parities. In the 40 years between 1939–40 and 1979–80, for example, the proportion that fourth or higher-parity births were of all births declined by some 40–90 per cent. At the beginning of the period, births at these upper parities ranged between 18 and 46 per cent of the total (with a median of 25 per cent); by the end of the period, this range had narrowed to but 3 to 16 per cent (with a median of 6 per cent) (Table 1.1 and Chart 1.2).

Table 1.1 Fourth and higher-parity births as a proportion of all births: specified low-fertility populations, 1939–40 and 1979–80

Country	Approximate period 1939–40	1979–80	% Decline
Australia	20	8	59
Belgium	22	7	68
Canada	36	6	82
Czechoslovakia	23	6*	72
Denmark	23	4*	81
Finland	29*	6*	81
France	26*	7	75
Germany (East & West)	20	5	75
Greece	39	5	87
Hungary	30*	5*	82
Italy	36*	9*	76
Netherlands	36	6*	84
New Zealand	18	10	41
Norway	22	6	85
Portugal	46*	13*	72
Romania	40*	16	61
Switzerland	23	5	79
United Kingdom†	19*	7	63
U.S.A.	27	10	63

* Birth order based on sum of live births and stillbirths.
† England and Wales only.
Sources: Calculated from data in U.N. 1949–50, Table 21, and U.N. 1981, Table 25.

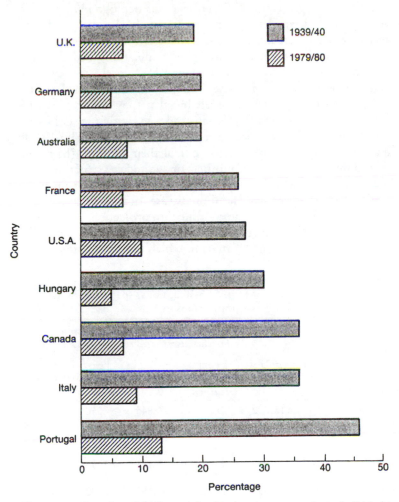

Chart 1.2 Fourth and higher-parity births as a proportion of all births, 1939–40 and 1979–80: selected low-fertility countries.

FUTURE POPULATION SIZE AND AGE STRUCTURE

What does this imply about future population sizes and age structures? The projections of both the United Nations and the World Bank are for these national fertilities to return to exact replacement levels (or to decline to these levels in the case of those few still above them) early next century. But this is only one of

5

several possible projections. To enlarge our demographic perspective on the matter, let us see what would happen under an alternative set of assumptions; one that, from the standpoint of those who are concerned about aging populations and declining numbers, may be viewed as something of a 'worst case scenario.' Let us assume for this purpose not that these national fertilities eventually reach replacement levels but that they continue at their 1985 below-replacement levels, instead. As with the U.N. and World Bank projections, let us also assume: (a) no decline in the capacity of these populations to control their fertility, (b) no net effect of migration on either numbers or age and sex structures, (c) a slight improvement in life expectation, and (d) no major mortality increases as a result of war, famine, disease, or natural disaster.

The eventual outcome of any population projection based on the continuation of a particular set of birth and death rates is a 'stable' population structure; one, that is, in which the proportions in the

Table 1.2 National population sizes at onset of near-stable* population structure under two sets of fertility assumptions: specified low-fertility populations

Country	1985 population (000)	World Bank fertility assumption		Continuation of 1985 fertility	
		Year† stability attained	Population (000)	Year† stability attained	Population (000)
Australia	15,752	2035	20,845	2050	19,869
Austria	7,555	2025	7,356	2060	5,098
Canada	25,379	2080	29,241	2055	24,486
Czechoslovakia	15,493	2100	18,386	2085	17,917
Finland	4,908	2015	5,171	2055	4,116
France	55,172	2025	61,806	2045	56,696
Germany (E & W)	77,659	2080	60,153	2080	31,858
Italy	57,128	2085	50,396	2065	34,684
New Zealand	3,254	2080	4,087	2060	3,712
Romania	22,740	2100	27,314	2085	27,271
Sweden	8,350	2085	7,405	2035	7,229
United Kingdom	56,543	2080	54,809	2060	46,033
U.S.A.	239,283	2025	286,181	2050	254,814

* Assumed to be when the proportion of the population age 65 and over is no more than half a percentage point below what it is projected to be in the year 2160.
† The nearest fifth year to the date when the proportion of the population age 65 and over is no more than half a percentage point below what it is projected to be in the year 2160.
Source: Calculated from data in World Bank 1986 and Zachariah & Vu 1988.

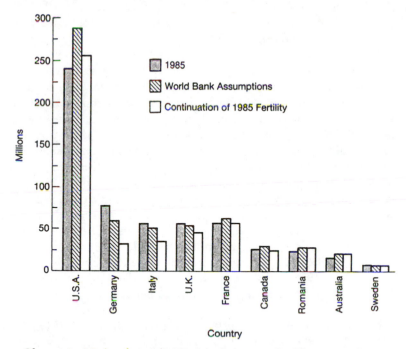

Chart 1.3   National population sizes in 1985 and at the onset of near-stable population structure under two sets of fertility assumptions: selected low fertility countries.

respective age/sex categories undergo no further change. A population with fertility below replacement level does not keep growing ever older. At a fixed set of age-specific fertility and mortality rates, it typically takes no more than two or three generations to produce a stable age structure (Preston, Himes & Eggers 1989, 692). Of course, in the present instance, because of the lower assumption about fertility, the population sizes emanating from our alternative assumption would be smaller than those projected by the U.N. or World Bank – and they would continue to grow still smaller because the fertility levels assumed for our purposes are not only low but actually less than what would be needed for replacement. Nevertheless, the result of each of these projections would be a 'stable' age/sex structure; not because of any behavioral constraints, but because of mathematical necessity: because, that is, they have been designed that way. Table 1.2 compares the results of our alternative projections with those of the World Bank. A selection of these comparisons is also presented in Chart 1.3. Like all such projections, the purpose of these is illustrative, not predictive.

As already noted, because they are based on the assumption of a continuation of below-replacement-level fertilities, the populations produced by our alternative projections would continue to decline even after reaching a stable age/sex structure. But one thing to note in this 'worst case' projection is that, even with continuation of these unprecedentedly low fertility levels, it would be many years before the numerical declines experienced by any of these populations would be sufficient to reduce their numbers to what they were on the eve of World War II – a level arguably sufficient for any legitimate national purpose, including the maintenance of an identifiable national culture and lifestyle. From the 1985 base year, it would, for example, require as many as 134 years (in France) and no fewer than 45 years (in Austria) (Table 1.3 and Chart 1.4). The extent of numerical decline a population would have to undergo to reach these 1939–40 levels ranges from a low of 12 per cent (Austria) to a high of 39 per cent (in both the Netherlands and the Soviet Union). The median decline would be 26 per cent.

Table 1.3 Population about 1939–40 and 1985, with number of years required with continuation of 1985 fertility levels to reduce 1985 total to that of 1939–40: selected low-fertility populations

Country	Population (000) 1939–40*	Population (000) 1985†	Assuming continuation of 1985 fertility	
			Year 1939–40 total would be reached‡	No. years to reach 1939–40 total‡
Austria	6,658	7,555	2030	45
Belgium	8,391	9,857	2034	49
Denmark	3,805	5,114	2039	54
Finland	3,686	4,908	2072	87
France	41,300	55,172	2119	134
Germany (E & W)	56,042	77,659	2037	52
Hungary	9,217	10,649	2033	48
Italy	43,112	57,128	2046	61
Netherlands	8,781	14,486	2069	84
Norway	2,954	4,153	2069	84
Sweden	6,326	8,350	2053	68
Switzerland	4,206	6,458	2062	77
United Kingdom	47,762	56,543	2053	68

Sources: * U.N. 1949–50, Table 3.
† World Bank 1986.
‡ Calculated from data in World Bank 1986.

OLDER AGE STRUCTURES

Further aging in these populations will occur, whatever happens to fertility. But if current (that is, 1985) fertility levels were to persist, increases in the proportions of old people would be especially marked. It is this change that seems to be the particular concern of those most worried about present demographic trends in these countries. What would this change amount to? Some illustrative projections are presented in Table 1.4.

If national fertilities rise to replacement levels and remain there – that is, if the World bank projections are realized – the proportions aged 65–74 will, in each of these populations, level off at about 10.4

Table 1.4 Proportion aged in stable populations* on basis of two different fertility assumptions: selected low-fertility populations

Country	% 65–74			% 75+		
	World Bank fertility projection	Constant 1985 fertility	diff.	World Bank fertility projection	Constant 1985 fertility	diff.
Australia	10.41	11.06	0.65	9.30	10.19	0.89
Austria	10.44	13.69	3.25	9.25	14.14	4.89
Belgium	10.44	13.58	3.14	9.27	14.00	4.73
Bulgaria	10.42	10.79	0.37	9.28	9.77	0.49
Canada	10.45	12.84	2.39	9.28	12.73	3.45
Czechoslovakia	10.39	10.50	0.11	9.29	9.37	0.08
Denmark	10.43	14.84	4.41	9.26	16.34	7.08
Finland	10.45	12.48	2.03	9.27	12.36	3.09
France	10.43	11.80	1.37	9.30	11.13	1.83
Greece	10.42	11.37	0.95	9.28	10.57	1.29
Hungary	10.43	12.85	2.42	9.27	12.85	3.58
Italy	10.36	14.15	3.79	9.33	14.97	5.64
New Zealand	10.45	11.04	0.59	9.23	10.40	1.17
Norway	10.42	13.19	2.77	9.28	13.29	4.01
Portugal	10.41	11.29	0.88	9.30	10.49	1.19
Romania	10.41	10.51	0.10	9.24	9.36	0.12
Spain	10.42	11.05	0.63	9.29	10.08	0.79
Sweden	10.37	12.97	2.60	9.32	12.94	3.62
Switzerland	10.44	13.61	3.17	9.28	14.05	4.77
United Kingdom	10.44	12.35	1.91	9.24	11.92	2.68
U.S.A.	10.41	12.85	1.44	9.30	11.23	1.93
Yugoslavia	10.41	10.84	0.43	9.27	9.86	0.59

Source: Calculated from data in World Bank 1986 and Zachariah & Vu 1988.
* Stable population assumed to be that population in the nearest fifth year when no subsequent change in the percentages in age groupings 0–14, 15–64, and 65 and over exceeds 0.05 per cent.

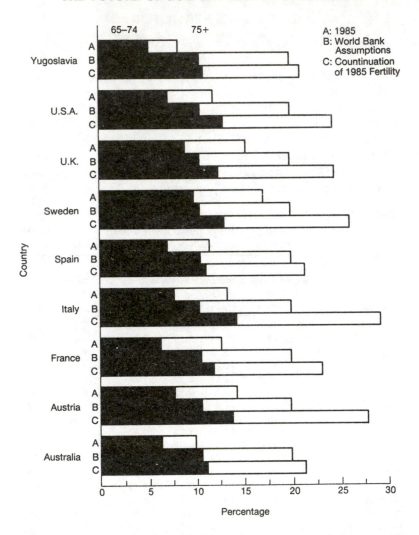

Chart 1.4 Proportion of population aged 65–74 years and 75+ years, in 1985 and in stable population derived from two different fertility assumptions: selected low-fertility countries

per cent; and those aged 75 and over, at about 9.3 per cent. This would add up to a total of slightly less than 20 per cent in the age grouping 65 and over (Table 1.4 and Chart 1.4). In comparison with the situation in the 1985 base year, this would entail increases in the proportion aged 65–74 ranging from as little as 0.58 of a percentage point (Sweden) and 1.55 percentage points (U.K.) to as much as 5.44 percentage points (Yugoslavia) (median = 3.42 percentage points). For the proportion aged 75 and over, it would entail increases ranging from a low of 2.25 percentage points (Sweden) to a high of 6.09 percentage points (Yugoslavia) (median = 3.95).

The aged proportions would be higher still if the alternative, 'worst case scenario,' projections were realized, instead, (Table 1.4 and Chart 1.4): those 65–74 leveling off at between a low of 10.5 per cent (Czechoslovakia and Romania) and a high of 14.8 per cent (Denmark) (median = 12.4 per cent); and those 75 and over leveling off at between 9.4 per cent (Czechoslovakia and Romania, again) and 16.3 per cent (Denmark, again) (median = 11.6 per cent). Against the World Bank projections, the alternative projection would add to the proportions in these age groupings amounts ranging at ages 65–74 from as little as 0.10 and 0.11 of a percentage point (Romania and Czechoslovakia, respectively) to as much as 4.41 percentage points (Denmark) (median = 1.68); and at ages 75 and over from as little as 0.08 and 0.12 percentage points (Czechoslovakia and Romania, respectively) to as much as 7.08 and 5.64 percentage points (Denmark and Italy, respectively) (median = 2.31). The additional amounts given rise to by the alternative projection would in each country be less at ages 65–74 than at 75 and over.

SEX RATIOS

It is sometimes forgotten that because women, on average, live longer than men, an older age structure will be less masculine. This could be expected to have important repercussions for both a population's living arrangements and the particular needs of its elderly sector. The recent declines in upper-age mortality rates intensify this condition because they have tended to benefit female survivorship much more than male. Particularly has this been so at the uppermost ages. In general, female life expectancies are consistently higher than male. But over the last decade or two, in the 20 of these populations for which comparative life table values are available, the proportionate gains in life expectancy at the older ages among women relative to men have tended to be greatest at age 85 (median = a 67 per cent increase in the ratio of female to male life expectancies) and least at age 65 (median = a 26 per cent increase), with the gains at age 75 (median = a 52 per cent increase) generally lying between these two extremes.

In Finland, for example, the excess in female over male life expectancy increased (between 1966–70 and 1985) by 32 per cent at age 65, 94 per cent at age 75, and 350 per cent at age 85. In the Netherlands, the corresponding increases (between 1972 and 1984–5) were 38, 94, and 118 per cent; and in Australia (between 1960–2 and 1985), they were 4, 26, and 41 per cent. Most of the rest of these populations experienced a similar pattern of increases, but not all: there were *declines* in the female-male differential in both the U.K. and the U.S.A. at age 65, and in Spain at age 85. Nonetheless, all of them retain the pattern of greater female life expectancy: the female excess ranging between 14 per cent (Greece) and 33 per cent (Finland and the Netherlands) at age 65, between 11 per cent (Bulgaria) and 38 per cent (Norway) at age 75, and between 6 per cent (Spain) and 28 per cent (England and Wales) at age 85 (calculated from data in U.N. 1973, Table 18, and U.N. 1986, Table 16).

GEOGRAPHIC DISTRIBUTION OF THE AGED

Each of these countries has regional differences in age structure – often quite marked ones. The way the regional boundaries are drawn, and the population sizes of the individual units these boundaries enclose, makes any precise international comparison impossible, but such of these differences as relate specifically to the elderly seem to be more extensive in Britain and the U.S.A. than elsewhere (see, e.g., Bohland & Treps 1982, Hugo & Wood 1984, Warnes & Law 1984, Kostrubiec 1987, Noin 1987, Parant 1987a, 1987b, Warnes 1987, Rudd 1989). The demographic origins of these regional differences lie in differences in fertility, mortality, and migration. These can exist simultaneously on both sides of the age range: a greater concentration in the upper ages, for example, can result from the out-migration of younger people as well as from the in-migration of older people, from declines in fertility as well as from increases in upper-age life expectations.

When it comes to assessing the need for services to the aged, it is, of course, their number and not their proportion in any particular area that demands the closer scrutiny. Hugo and Wood (1984, 52), for example, in studying local government areas in Australia, found that, while the *proportions* of older people were higher in resort–retirement areas, their *numbers* were higher in the major provincial urban centers. And yet, the proportion of older people in a locality's population can itself be a matter of some significance. For one thing, it will have implications for the dependency burden with respect to local financial arrangements and the provision of certain services. Moreover, regional polarization of the aged – with particularly high concentrations in some regions and particularly low concentrations in others – can affect the outcome of elections (to the extent that the aged follow a different pattern of voting) as well as the frequency and intensity of older people's contacts with kin and friends and the range and quality of the tasks these people – the elderly and their kin and friends – will be able to perform for one another. To the extent that social participation is defined by age, it is also possible that the risk of social isolation for older people – at least in modern industrial societies – will be greater where they are a smaller minority than where they are a larger proportion of the total – if only because, with the former, they might not constitute a critical mass sufficient to ensure either the environmental conditions or the personal consideration appropriate to their interests.

Then there is the matter of 'visibility.' A society's awareness of its aged members is a function of much more than their proportion of the population (it is a function, also, of their incomes, conditions of health, and types of housing, for example), but it is reasonable to suppose that this awareness will be greater where the aged are a larger share of the total. Marked regional differences in age structure – by geographically concentrating the elderly – could have the effect of reducing their visibility in society as a whole, with a discounting of their needs and interests on the part of officials and the general public as a possible consequence. While such concentration seems capable of increasing potential political strength in a few electorates, it also seems capable of removing a large portion of the aged from that degree of participation in the larger society that would seem to be desirable from the standpoint of both their own wellbeing and the wellbeing of society as a whole. As with so much else, there are both costs and benefits; and what is good for the individual at one particular time might not be so good for the society – or for the individual at some other time.

REFERENCES

Bohland, J. R. & L. Treps 1982. County patterns of elderly migration in the United States. In *Geographical perspectives on the elderly*, A. M. Warnes (ed.), Chichester: John Wiley.

Carr-Saunders, A. M. 1936. *World population*. Oxford: Oxford University Press.

Durand, J. D. 1967. The modern expansion of world population. *Proceedings of the American Philosophical Society* 111(3), 136–59.

Hugo, G. & D. Wood 1984. *Ageing of the Australian population: changing distribution and characteristics of the aged population*. 1981 Census Project Paper 8, Bedford Park, South Australia: Flinders University, National Institute of Labour Studies, Working Paper Series No. 63.

Kostrubiec, B. 1987. Evolution du processus de vieillissement de la population polonaise. *Espace, Populations, Sociétés*, no. 2, 329–42.

Noin, D. 1987. La population très âgée en France. *Espace, Populations, Sociétés*, no. 1, 29–40.

Parant, A. 1987a. Le vieillissement démographique des cantons français de 1962 à 1982. *Espace, Populations, Sociétés*, no. 1, 75–86.

Parant, A. 1987b. Le vieillissement démographique: un nouveau défi pour le Japon. *Espace, Populations, Sociétés*, no. 2, 329–42.

Preston, S. H., C. Himes & M. Eggers 1989. Demographic conditions responsible for population aging. *Demography* 26(4), 691–704.

Rudd, D. 1989. *The ageing of local area populations in Victoria: past patterns and projected trends in the aged population*. Prepared for the Aged Care Research Group. Carlton, Vic., LaTrobe University, mimeo, 151 pp.

U.N. (United Nations) 1949/50, 1954, 1973, 1981, 1986. *Demographic yearbook*. New York.

Warnes, A. M. 1987. The distribution of the elderly population of Great Britain. *Espace, Populations, Sociétés*, no. 1, 41–56.

Warnes, A. M. & C. M. Law 1984. The elderly population of Great Britain: locational trends and policy implications. Institute of British Geographers: *Transactions*, 9(1), 37–59.

World Bank 1986. World population projections. Unpublished, Washington, DC.

Zachariah, K. C. & M. T. Vu 1988. *World population projections, 1987–88 edition. Short-and-long-term estimates*. Baltimore, Md: Johns Hopkins University Press.

2

THE FUTURE OF FERTILITY
AND MORTALITY

However we choose to define old age, the changes that can be expected in age structures and growth rates in these countries are bound to occasion some difficulties. But in so far as they are the direct result of older age structures and declines in growth rates (or of declining numbers themselves), these difficulties may be less serious and less intractable than is commonly envisaged. For one thing, the demographic changes out of which they emanate will not occur suddenly. For another, there may be less 'deterioration' in the demographic situation itself than is commonly expected. There are two demographic processes to be considered: fertility and mortality. The role of migration will be dealt with later.

THE RATE OF PROGRESSION TO NUMERICAL DECLINES AND OLDER AGE STRUCTURES

Considered in terms of the broad sweep of human history, the expected changes in population size and age structure will be rapid and extensive. Considered in terms of individual human lives, however, they will be gradual and all but imperceptible. To put some perspective on the matter, let us again employ that 'worst case scenario' of Chapter 1, making some projections on the assumption of a continuation of those unprecedentedly low, below-replacement-level fertility rates of 1985. Some results of these projections for relative population size and the demographic prominence of the elderly are presented in Table 2.1.

On present low-mortality levels, these projections encompass the individual lifetimes of nearly everyone alive today. Were they to eventuate, there would be in the United Kingdom, for example, a *decrease* of 9 per cent in total numbers by the time half of the

15

Table 2.1 Indexes of projected relative population sizes, and of numbers and proportions 75+ years of age in specified years, assuming continuation of 1985 age-specific fertility rates: selected countries.

	Year						
	1985	2000	2015	2030	2045	2060	2075
France							
Population	100	106	109	108	103	97	91
Persons 75+	100	102	120	159	177	169	161
% 75+	100	96	111	147	172	175	177
Germany (East & West)							
Population	100	97	89	78	65	53	44
Persons 75+	100	94	121	126	152	117	102
% 75+	100	97	136	162	234	219	234
Italy							
Population	100	101	96	88	77	64	54
Persons 75+	100	123	152	167	197	168	146
% 75+	100	122	158	190	257	261	270
Sweden							
Population	100	99	96	89	81	71	63
Persons 75+	100	114	109	143	145	128	117
% 75+	100	115	114	160	181	179	184
United Kingdom							
Population	100	100	99	95	88	81	75
Persons 75+	100	112	110	133	160	146	143
% 75+	100	111	112	140	182	179	191
U.S.A.							
Population	100	109	115	115	109	102	97
Persons 75+	100	127	137	223	261	247	233
% 75+	100	116	120	195	240	242	243

Source: Calculated from data in World Bank 1986 and U.N. 1985, Table 36.

20-year-old men in that country in 1985 had died, and, with respect to persons 75+ years of age, *increases* of a half (53 per cent) in number and two-thirds (68 per cent) in their proportion of the total population – or, in the latter instance, an increase of 4.2 percentage points to 10.5 per cent from the 6.3 it was in 1985. Where numerical declines occurred (and because of the national differences in historical fertility patterns they would not occur in every instance), the extent of the change experienced by each successively older cohort in 1985 would be progressively less. By way of contrast with the experience of the 20-year-olds, the surviving half of men who were 60 years of age would encounter a population whose size was equal to what it had been in 1985, and in which the increases in the number 75 and over and the percentage these were of the total population would have been but 12 and 11 per cent, respectively. To adjust to the more extensive changes they would encounter, the surviving half of the cohort of 20-year-old men would have had almost 55 years (from 1985 to mid-way between 2039 and 2040); to adjust to the less extensive changes, the cohort of 60-year-old men would have had just under 17 years. These and comparable projections for other populations are shown in Table 2.2.

In some countries – those, like France, Sweden, and the U.K., in which low fertility has had a longer history and the post-World War II fertility increases were less pronounced – the experience of these

Table 2.2 Projected extent of changes in total population and in population 75+ years of age experienced by time half of men of specified age in 1985 will have died, assuming continuation of 1985 age-specific fertility rates: selected countries.

			Persons 75+	
	1985 age	Total population % decline (increase)	% increase (decrease) in total number	% increase (decrease) in per cent share of pop.
France	20	(5)	74	66
	40	(9)	22	12
	60	(7)	8	(4)
Germany (E & W)	20	30	47	89
	40	14	23	44
	60	4	(3)	1
Italy	20	19	88	131
	40	6	52	62
	60	0	21	21
Sweden	20	18	44	75
	40	7	37	48
	60	2	17	19
U.K.	20	9	53	68
	40	2	15	17
	60	0	12	11
U.S.A.	20	(9)	165	139
	40	(15)	58	37
	60	(11)	32	19

Source: Calculated from data in World Bank 1986 and U.N. 1985, Table 36.

demographic changes on the part of persons now alive will be relatively slight. In other countries – those, like the U.S.A., with more extensive fertility increases following upon World War II, or, like Italy, with shorter histories for both low fertility and low mortality (Caselli & Vallin 1990) – it will be greater. But even in these latter countries, the demographic progression to smaller numbers and older age structures – at the level of individual experience, at least – will be gradual; certainly it will be nothing like the post-World War II 'baby booms' that, within but a few years, not only brought marked numerical increases to these countries but introduced sizable distortions into their age structures as well. In the U.S.A., for example, the more elevated birthrates of the 1947–59 period added almost a quarter (23 per cent) more births to the population than would have been added had the still relatively high birthrates of the 1940–5 period remained in force, instead. The marked distortion in the age structure this occasioned is epitomized in the fact that the birth cohort of 1947 was more than a third (35 per cent) larger than that of those born only two years earlier (calculated from data in U.N. 1949–50, Table 15, 1959, Table 9, and 1965, Tables 11 and 12). While death can take place over the entire range of ages, the occurrence of birth is necessarily restricted to age 0.

THE EXTENT OF DEMOGRAPHIC 'DETERIORATION'

Any increase in growth rates, whether from an excess of births over deaths or of immigrants over emigrants, will slow the rate of the transition to smaller numbers and older age structures. Numbers will decline more slowly, perhaps stabilize, possibly even increase; while the transition to an older age structure will simultaneously proceed more slowly. For present purposes, it is enough to note that the consequences of migration in this respect, which will be discussed in Chapter 5, are less certain than those of natural increase (that is, than those of an excess of births over deaths).

Fertility

Most demographers have assumed that control over fertility would make annual birthrates (but not necessarily ultimate fertility levels themselves) more responsive to 'transitory phenomena,' like recessions and political events. 'The future of fertility,' wrote a prominent demographer over two decades ago, 'is likely to be increasingly bound up with questions of fluctuation rather than trend. ... My impression is that most of [the recent drop in the birthrate] has transitory origins' (Ryder 1969, 116). Voicing a similar view, another prominent demographer has predicted that the more developed nations 'will have an endless series of ups and downs in their births' (Keyfitz 1972, 361). Although reasonable enough on the basis of experience up to the time they were made, such predictions have not been realized in the period since. This would seem to be because of certain changes in the countries in question: changes of a sort likely to make fertility less affected by the types of social and economic factors or fads that have prompted fluctuations in annual numbers of births in the past. Each of these countries has witnessed the development of a greater openness and frankness about human sexuality, a wider range of roles for women alternative to those of mother and homemaker, changes in attitudes and tastes relating to what is considered an 'appropriate' family size, and a downward narrowing of the range of family sizes so that there are fewer examples of higher-parity 'deviancy' to serve as encouragements to similar levels of procreation among those otherwise unlikely to embark upon them.

At the same time, these countries have also experienced improvements in contraceptive technology and, most of them, increases in the availability of birth control techniques (including legal abortion), as well. These latter developments could, of course, result in greater, rather than less, short-term fluctuation in the birthrate, as couples changed the timing of their births in response to fluctuations in, say, employment levels. But in association with the broader changes that have occurred in institutions and individual values, the improvements realized in birth control technique and availability seem to have led, instead, to more rather than less, procreative stability – certainly in ultimate family size and possibly, also, in the timing of childbearing.

Still, the possibility of either further declines or subsequent increases in fertility should not be ruled out. The gap of some four-fifths of a child between the average number of children a woman would bear under continuation of present fertility patterns in such low-fertility countries as Poland and Romania, on the one hand, and Germany and Denmark, on the other (U.N. 1986, Table 22), suggests at least the possibility of some further movement in either direction. Further declines are suggested by the fertility trends of the past couple of decades, which have seen the highest rates of decline, in both absolute and proportionate terms, taking place in those populations with the highest fertility levels at the outset.

However, in none of these countries does fertility seem likely to drop below the levels already attained by those countries with the lowest fertility levels. And it is at least arguable that increases in the fertilities of these populations are more likely than are further declines – certainly increases to levels higher than those now current, if not necessarily to levels sufficient for full replacement. To the extent increases occur, demographic 'deterioration' in these countries will be less than that predicated on the basis of present patterns. If, for example, the World Bank's projections of fertility rising to exact replacement level by the year 2020 are realized, the proportion of the population aged 75 and over in Denmark in the year 2050 will be 12.1 per cent as against 15.9 per cent with continuation of 1985 fertility levels.

The very fact that these countries' mid-1980s fertility levels were so low – in most of them the lowest ever – itself argues for at least some increases (Ermisch 1981). Nor would these be unprecedented. Marked increases in fertility following upon periods when it was very low have occurred in a number of these countries (although, admittedly, the times – of depression and war – were notably atypical and the low fertility levels in question had not descended to quite the depths of the mid-1980s or remained, comparably, so low for so long). Moreover, the slight increases since the mid-1980s in total fertility rates in, for example, Sweden, Germany, Norway (I.N.E.D. 1989, Table 16), and the U.S.A. (Ahlburg & Vaupel 1990, 642) as well as in parity progression ratios (that is, in the proportions with n children going on to have another child), suggest the imminence of at least some fertility increases. There is, however, no suggestion in any of this that the fertility levels of these countries will soon rise to replacement levels; only that there will be some increase in fertility and, therefore, some slowing down of the progression toward that demographic 'deterioration' seen as consisting of older age structures and numerical declines.

But even in the absence of evidence that something of the sort may actually be taking place, an argument for some future increases in fertility would not seem altogether out of place. There is, for example, the possibility of a renewed emphasis upon family and parenthood, with some further encouragement to childbearing as a result. In the past, divorce has tended to reduce fertility levels. But the fact that, controlling for women's ages, the average numbers of children ever born among American women who have been married more than once have come increasingly to equal those among women married only once (and living with their husbands at the time of the census) suggests that divorce may now be tending to increase fertility, instead. Women (and men) who, had they remained in their first unions, might otherwise have ceased reproduction after one or two births, may, upon dissolution of the one union and formation of another, come increasingly to view the birth of a child or two to the new union as something of a cementing factor as well as a tangible expression of love and success following upon the failure of the earlier union.

Moreover, greater emotional value may attach to children as larger numbers of men and women find they receive insufficient psychic gain or emotional support from job, neighborhood, kin, or religion; or as they respond to a more acutely perceived lack of control over their destinies by withdrawing further into the bosom of the nuclear family. As Demeny has written semi-seriously (1987, 352), 'In a Herman Kahn-like exurbanized super-affluent future, the tedium of working at home at one's computer terminal may be relieved by rediscovery of the fun of having children around.'

Nor is it inconceivable that the participation of women in the labor force (which has been widely associated in these countries, either directly or indirectly, with lower fertility) might decline somewhat in favor of a renewed emphasis on the traditional roles of mother and homemaker. Women have this alternative role available to them; men do not. A number of changes, more in perception than in actual fact, could evoke such a response. There could, for instance, be a growing impatience with the continuation of unequal economic treatment *vis-à-vis* men; the sort manifested, for example, in lower relative wage rates and fewer opportunities for promotion. Whatever the reasons for its continuation – and the decline in membership and power among labor unions as a result of changes in occupational and industrial composition, government opposition, and worker apathy and dissatisfaction over union leadership seems an important element – such inequality is hardly conducive to satisfaction among a generation of women acquiring more education and training and taking to heart current encouragements to assume more economic and personal initiative.

There could also be a growing dissatisfaction with working

outside the home and, in particular, with the sorts of living conditions – the time-centeredness, the commuting, the lack of time for oneself or for being with one's child – this imposes.

Still another possibility is the development of a lesser receptivity to the appeals of consumerism. This could come about in several ways. As a result of changes in social policy, some services or experiences that were formerly available, if at all, only through private purchase could become available at much lower, even zero, cost. Parks, beaches, outdoor recreation facilities that formerly required for their enjoyment a private automobile and days off work (paid or unpaid) would fall into this category. So would medical care, care in old age, and higher levels of schooling for oneself or one's family, as well as some elements of entertainment and home maintenance. Given its particular significance as an item of personal expenditure in the countries under consideration, anything that permits substitution of public for private transportation (or that lessened the necessity – or desire – to travel) could significantly reduce the need for income and, hence, the desire of married women to enter the labor force.

Nor is satiety an impossibility. Among the many incentives (social, psychic, and economic) to women's participation in the labor force, the desire for money to buy not the bare essentials of life but the multiplicity of things beyond these essentials that one has learned to want is surely among the most potent. The insatiability of human wants may be something of an axiom in the economist's frame of reference, but that does not make it true. Advertising, that quintessential industry of the growth economy (Potter 1954, Ch. 8) – devoted, as it is, to the creation of wants for an ever-expanding supply of products, takes the possibility of satiety as a given – and then proceeds on the assumption that, with skill (and money, and sometimes bombast), it can be sufficiently overcome to ensure a profit. In this particular instance, advertisers appear to understand human nature and behavior better than economists.

There could also be less pressure to buy, as the result of a lower rate at which wants were created. This could come about in several ways, pressures to buy being both direct and indirect, concentrated and diffuse. Basically, it would involve a lessened exposure to advertising (in consequence of environmental or governmental limitations on it, for example) and, related to this, reduced buying on the part of one's peers and, in particular, on the part of one's 'significant others' (Gerth & Mills 1954, especially Ch. 4).

Finally, there could be a disappearance or at least a decline in the availability of certain goods, sights, experiences for the purchase (and presumably the enjoyment) of which people have shown themselves willing to work and save. In this category one could list, for instance: (a) goods made unavailable in consequence of, say, controls on behalf of environmental quality; (b) the possibility, lost forever as a result of urban development or the construction of roads and shopping centers, of having a vacation house or a larger plot of land, and (c) sights and experiences (expecially of a touristic nature and in other locales) effectively ruined or destroyed by overuse or overdevelopment, or by the encroachment of population growth or industrialization.

It is, of course, possible that these developments will not take place; or that they will not take place at a rate or to an extent sufficient either to evoke any marked increase in fertility or to overcome such counterpressures against childbearing as are to be found, nowadays, in the likes of housing shortages, money worries, the lack of childcare facilities, uncertainty about the future (with respect to war, environmental quality, or politics, for instance), interest in immediate gratification, or the unwillingness to commit oneself. But they should not, at least for the present, be ruled out of the range of possibility so far as their potential influence on fertility decision-making is concerned. Just as the present rates at which people are increasing in both numbers and consumption levels must necessarily be either a brief interlude in the long course of human history or else humanity's last, final gasp, so also must much of what has been distributed through the market, and therefore through the purchase by individuals with the money to buy it, either cease to exist or become no longer so desirable that it offers sufficient incentive for people (especially those who have some socially acceptable alternative) to sacrifice their time, energies, and freedom of movement in such disciplined pursuit of it as, at present, characterizes so much of the participation in the labor force. It would be another matter if this participation were to become less demanding; but were this to happen, participation in the labor force would presumably be more amenable to being combined with childbearing.

Mortality

Current forecasts of longevity among the populations under consideration are unanimously for further increases, and, thus, for still older age structures. But, first, a word about the concept of old age itself. There is more than one way to look at it. The character Jaques in Shakespeare's *As You Like It* sees the life course as a succession of roles in a drama:

> All the world's a stage,
> And all the men and women merely players.
> They have their exits and their entrances,
> And one man in his time plays many parts,

These parts, in Jaques' reckoning, commence with the infant 'mewling and puking in the nurse's arms.' They end with the extreme of old age, which he characterized as a time of

> ... second childishness and mere oblivion,
> Sans teeth, sans eyes, sans taste, sans everything.
>   (Act II, Scene VII)

Age-grading is a universal of human society. The more formal types – like minimum ages for leaving school, driving a car, seeking employment, marrying, drinking alcoholic beverages, receipt of an age pension – are defended as requisite to administrative efficiency and human wellbeing; the less formal – the typing of people as 'children,' 'adolescents,' 'adults,' 'old people' – are attributed to biology and human nature. Arbitrariness in categorization is basic to both, and neither is as poetic as Jaques'.

The strictly demographic approach defines age in terms of birthdays and then neatly categorizes the population accordingly. It fits in with most administrative needs and permits a variety of actuarial calculations and projections of particular use to modern states and industrialized economies. But it is an approach that says nothing about the condition of the people at each successive age level; nothing about their physiological, mental/emotional, or economic condition. And it only hints at their social positions and the ways they might be performing the roles associated with these positions; the ways, in a more general sense, in which they might be spending their time.

It is this failure to look beyond birthdays, this failure to consider human beings in all their glorious heterogeneity and complexity, that lies at the root of much of the disquiet about current demographic developments in low-fertility populations. In consequence of (a) greater longevity at the upper ages and, in particular, of (b) what, for the most part, are both the lowest fertility levels in their histories and their longest-ever periods of below-replacement level

24

fertility, the age structures of European populations are becoming increasingly more concentrated in the upper ranges. By the middle of the 1980s, the proportion of the population 65 years of age and over had risen in the United States, for example, from the 4 per cent it was at the beginning of the century to 12 per cent; in France, from 8 per cent to 13 per cent; in England and Wales, from 5 per cent to 15 per cent; and in Sweden, from 8 per cent to 17 per cent (calculated from data in Taeuber & Taeuber 1958, Keyfitz & Flieger 1968; World Bank 1986). The concern this has given rise to is but heightened by the fact that it is the highest age categories that have, in recent years, experienced the greatest increases. While increases in the proportions aged 65–74 over this period were in the range of some 10 to 170 per cent, those in the proportions aged 75 and over were in the range of some 160 to 360 per cent (calculated from data in Keyfitz & Flieger 1968, and World Bank 1986).

Arbitrariness in categorization by age is particularly pronounced with the elderly. Not only are they both socially and economically the most heterogeneous sector of the population, they are also highly various physiologically and emotionally (Butler 1975, 7). Precise determination of the onset of the particular characteristics and capacities commonly ascribed to old age is impossible. (For that matter, so is precise determination of the consequences, whether for the individual or the society, of a person's attaining this stage of life.) It makes no difference whether the issue is addressed in terms of physiology or behavior. It is like this with the other categories based on chronological age, too; it is just that, with old age, there is far greater variation among those so categorized, and the variation that exists extends over a markedly wider age range.

The current disquiet about aging is, of course, based on age defined in terms solely of birthdays. The results are somewhat different if we reason, instead, in terms of life expectancy. Defining old age in the usual administrative way as beginning at chronological age 65 produces one set of parameters concerning the 'aged' proportion in the population. Defining it in terms of a different – although still arbitrary – type of demographic definition, namely, the proportion in those age categories where life expectancy is less than 10 years (see Ryder 1975) produces another. A life expectancy definition of old age is no less arbitrary than one based on birthdays, but it offers more insight into the physiological condition of the older population, and so may, for that reason, be more appropriate to considerations of social policy. Table 2.3 compares

Table 2.3 Proportion of the population classified as 'aged' at three points in time, by age* and life expectation† criteria: specified low-fertility countries.

Country	1940 Age	1940 Longevity	1985 Age	1985 Longevity	Future time when population returns to 1940 size Age	Future time when population returns to 1940 size Longevity
Denmark‡	7.4	5.1	12.8	5.9	26.5	18.0
Finland	6.4	4.6	12.3	8.3	24.6	18.2
France‡	10.3	8.3	10.0	6.2	20.4	14.3
Germany (E & W)	9.8	8.3	14.3	8.3	29.9	21.9
Hungary	7.0	4.9	12.6	7.7	20.6	18.1
Italy	7.5	5.3	13.2	6.4	29.4	22.0
Netherlands‡	6.7	4.6	9.9	5.2	26.4	19.7
Sweden‡	8.7	5.6	14.9	7.5	22.6	16.2
Switzerland‡	7.5	5.2	11.6	5.6	24.7	17.5
United Kingdom§	8.1	6.6	15.2	8.4	23.1	16.7

* Age criterion for 'aged' classification is age 65 or older.
† Longevity criterion is having a life expectancy of less than 10 years.
‡ Males only.
§ England and Wales only, in 1940.
Sources: Calculated from data in: World Bank 1986 and U.N. 1949–50, Table 4, and 1960, Table 5.

the results of the application of these two definitions to populations for which the data are available for three different periods: (a) the eve of World War II, (b) the present (that is, the 1985 base year), and (c) that hypothetical future when, if continued, present fertility levels will have reduced total numbers to what they were on the eve of World War II.

On the basis of the 65+ criterion, the proportion 'aged' on the eve of World War II (in the ten countries for which the calculation can be made) ranged from 11.4 (France) to 6.4 per cent (Finland). Applying the alternative – life expectancy – criterion reduces these percentages to 7.2 and 4.6 respectively. Similarly, application of the alternative criterion reduces the extent of the increase in the 'aged' proportion between the eve of World War II and the time when continuation of 1985 fertility levels will have reduced total numbers to what they were at that earlier time. Comparisons for total populations are limited because the highest age group in the World Bank projections is open-ended after age 75. But with those comparisons that can be made without estimating the age distributions within this highest age category, the reduction from the figure derived from application of the life expectancy criterion is on the

26

order of 25 to 30 per cent. For example, instead of the 29.9 per cent figure for the aged population in Germany yielded by the 65+ criterion, the life expectancy criterion yields a figure of 21.9 per cent; instead of 23.1 in the United Kingdom, it yields 16.7; instead of 29.4 in Italy, it yields 22.0; and for French males only, instead of the 20.4 yielded by the 65+ criterion, the life expectancy criterion yields 14.4 (calculated from data in Preston, Keyfitz & Schoen 1972, and U.N. 1949–50, Table 4, 1957, Table 24, 1960, Table 5, 1983, Tables 7 and 22, 1984, Table 22, 1985, Table 34). To be sure, aging has – and will have – taken place, but much less so by this arguably more socially and demographically meaningful criterion than by the type of criterion usually applied. If the life expectancy criterion does not eliminate 'aging,' it most certainly reduces it.

But it is also useful to look more broadly at old age itself. The consequences of old age are neither precise nor certain. However administratively useful it may be, for example, to plan or establish eligibility for programs for the aged in terms of birthdays, there is no direct causal relation between age and need. Much depends on habits of life prior to old age: on occupation, diet, exercise, the experience of stress and the intake of drugs, for instance. Much also depends on contemporary conditions, like air quality, noise, and the availability of public transportation and appropriate housing. At a time when armies were moved on foot and hand-to-hand combat was the norm, it was 'not unusual,' according to Kebric (1988), to find among centurions, those battlefield mainstays of the Roman legions, men who were in their 50s or even 60s. Many among both officers and foot soldiers in the army of Philip of Macedon (382–336 B.C.) and, later, in that of his son Alexander (356–323 B.C.), the first 'large, well-trained, well-led professional national army' in Europe, were well into their 60s and 70s – 'far beyond the age today's society generally considers useful for anything' (Kebric 1988, 300), let alone for foot-soldiering and hand-to-hand combat. One might reasonably expect some age succession of this sort to take place at a time when campaigns were apt to last long enough for the young to become middle-aged and the middle-aged to become old. But the examples from ancient times are not restricted to the military. Kebric notes that some among the Roman charioteers, a profession, in his view, probably unmatched for its dependence on 'reflex, skills and strength,' 'continued to race in the Great Circus even when they were senior citizens' (p. 302).

More recent examples are found in Benet's (1974) report on the

Abkasians of the Soviet Republic of Georgia, among whom unusually high proportions not only survive to very advanced old age but maintain physical and emotional fitness while doing so. Although not ruling out genetic factors altogether (p. 103), Benet explains this phenomenon primarily in terms of the cultural, social and psychological factors structuring the Abkasians' existence (pp. 103–7).

Short of accepting the alternative, old age cannot be prevented. But aging is in many ways as much a social and psychological process as it is a biological. *When* old age begins and *how* it manifests itself may well be explained less by considerations of birthdays than of conditions of life.

While the increase in the proportion 'aged' between the eve of World War II and the mid-1980s ranged between a low of −3 per cent (France) and a high of 92 per cent (Finland) (median = 71 per cent), if old age is defined as commencing at chronological age 65, the range of this increase was between lows of −25 per cent and 0 per cent (France and Germany, respectively) and a high of 80 per cent (Finland) (median = 21 per cent) if old age is defined, instead, as beginning when life expectancy is no more than 10 years (Day 1988).

Of course, whatever the fertility and mortality assumptions employed – and however one defines old age – any reasonable forecast must be for further increases in the proportions of the aged in each of these countries under consideration. Applying a life expectancy criterion may lessen the extent of the perceived aging to be expected, but it can hardly prevent aging itself.

Numbers and sex ratios are a function of the operation of three demographic variables: fertility, migration, and mortality. Movements in fertility and migration can have some effect on the *distribution* of the population by age (defined in terms of birthdays), but it takes many years before such movements can have much effect on the actual *numbers* of the elderly. If old age is defined as commencing at age 65, any direct effect on these numbers arising out of a change in fertility would not take place for another 65 years. So far as *migration* is concerned, given the usual concentration of migrants in the young adult ages, any direct effect on numbers of the aged would ordinarily not take place for another 20–45 years. Any contribution of migrants' *fertility* to the numbers of elderly would, of course, commence even later.

This leaves *mortality* as, essentially, the sole determinant, for another generation or two, of numbers and sex ratios among the elderly in the populations under consideration. So what can we expect by way of future mortality levels in these low-mortality populations? To forecast mortality (or, for that matter, nonlethal morbidity) is to forecast something about: (a) the nature of the agent initiating the particular condition leading to death (for example, the infective organism or the cause of accidental death), (b) the condition of the potential host (that is, the individual person) who might come into contact with the infective organism or experience the traumatic event), and (c) the condition of the environment – natural, social, and economic.

The initiating agent can become more or less widespread, thus affecting the likelihood of exposure; and it can become more or less virulent, thus affecting both the likelihood that contact will result in trauma or infection and the likelihood that any trauma or infection will be severe. The condition of the host will affect his or her ability to withstand infection or avoid trauma, and, if affected, to recover. The condition of the environment will affect the likelihood of disease or trauma, the likelihood of recovery among those experiencing disease or trauma, and the extent to which the afflicted or handicapped person will be able to cope with his or her condition.

The forecasts of further declines in mortality at the upper ages are based on one or the other of three assumptions: (1) the assumption of an almost mechanistic continuation of current downward trends in mortality at advanced ages – or a reassertion of the downward trends of former years where, as in some Eastern European countries (particularly among males), these trends have been stalled or reversed (for discussion of the situation in Hungary, see Compton 1985, and in Poland, Prominska 1990). It is as though the momentum of inertia were itself the cause, thus borrowing from physics not so much a metaphor as a theory of causation; (2) the assumption that there will be mortality-reducing improvements in the prevention of disease and trauma; and (3) the assumption that there will be mortality-reducing improvements in therapy.

These assumptions relate, in particular, to other assumptions about: (a) developments in science and its application through preventive and therapeutic medicine and rehabilitation; (b) improved access to medical and rehabilitative services, together with custodial care, through removal of economic, administrative, and geographic barriers that, at present, limit such access; and (c) changes in lifestyle of a sort that reduce the individual's exposure to disease or trauma and improve his or her ability both to withstand infection and to recover from infection and trauma.

Are such assumptions reasonable? On the basis of an analysis of 19th century morbidity and mortality data in the records of British insurance companies, Alter and Riley (1988) challenge the common assumption that rates of death and illness necessarily move in the same direction: the assumption, that is, that high or low rates in the one indicate, respectively, high or low rates in the other. Their data show a marked increase in morbidity in the 1840s–1890s – at precisely the time when mortality rates in the population they studied were falling. To account for this seeming paradox they employ two models: 'constant frailty' and 'insult accumulation.' The former assumes that excess mortality among its frailer members leaves a cohort composed of less frail persons and therefore of persons less susceptible to death; the latter, that the longer people live, the greater is their accumulation of physiological 'insult' through the experience of illness, and, therefore, the greater their susceptibility to disease. If public health and, especially, medical measures succeed in improving recovery from illness – and therefore bring about a lower death rate – without at the same time reducing either the number of episodes of ill health or the severity of these episodes, one consequence will be a cohort that has experienced more 'insult' – and that is, therefore, more susceptible to illness. Alter and Riley think this is what happened with the 19th century British population whose records they studied.

But science and medical practice have progressed well beyond the levels reached in the 19th century. Today's technology, Alter and Riley believe, is capable of reducing not only the number of episodes of disease but also the duration and severity of those episodes that occur. This offers for the first time, in their view, the possibility of combining longer lives with less – rather than more – illness.

It is a heartening assessment, and certainly one amply supported by the evidence – to date. But can we expect it to continue well into the future? Further developments in health-related science will doubtless occur, whatever the situation with respect to the application of these developments. The cumulative nature of science will see that they do. The main hope appears to lie in future gains in immunology – and the drug therapies these might give rise to. But current optimism about such developments is largely confined to: (a) diseases like malaria, which, however significant on a world scale, affect few in the populations under discussion here, or (b) conditions, like senile dementia, which result in great disability and cause great distress, but which do not kill, or (c) conditions whose direct consequences are visited largely upon infants and children rather than the elderly (Brent 1986, Nossal 1987).

Moreover, in assuming (a) further beneficial developments in science and rehabilitation, (b) improved access to services and care, and (c) a changeover to more healthful lifestyles, those who forecast further gains in health – or at least longevity – among the elderly are leaving out of account certain other factors that could have a counterbalancing effect.

One such factor is the possibility of change in the virulence of infective organisms. Within present adult lifetimes, there has been a marked decline in the virulence of, for example, scarlet fever and pneumonia (at least among the young and middle-aged). But will future changes in virulence necessarily follow a downward path? Past fluctuations in the virulence of such diseases as scarlet fever, influenza, syphilis, and possibly measles suggest that it may be more accurate to forecast both increases and decreases.

There is also the possibility of new diseases. AIDS is only a recent, and particularly notorious, example; the varieties of influenza are another.

There is the possibility of new epidemics as a result of the monoculture of domestic plants and animals (which increases their exposure to devastation), the increased mixing of human populations that comes with the growth of urbanization and the expansion of air travel between widely separated populations, and the improvements in sanitation and vaccination that 'leave the larger human herd more innocent of microbial experience' (Lederberg 1988).

Then there is the possibility of a more general worsening of environmental conditions: the depletion of ozone, the accumulation of toxic materials in agriculture, acid rain, heightened radioactivity, global warming, air and water pollution, ground water pollution, unhealthful chemical agents in food, for example (W.H.O. 1980–).

And there is the possibility of additional stress. Whether stress will be more intense and widespread in the future than in the present is impossible to say. But it would appear to be more intense and widespread now than it was a generation or two ago, what with recent trends toward, among other things: the concentration of population in large metropolitan centers, increases in armed civilian violence, greater uncertainty arising out of a greater variety of social roles (and a greater range of possible behaviors attached to these roles), increased geographic mobility, unemployment among new entrants into the labor market that is both more extensive and of longer duration, increased amounts of advertising leading to augmented wants and greater frustration at not being able to satisfy these wants. The list is long. Just what are the consequences of stress for health and wellbeing is, as yet, little known. It is at least possible that the experience of stress – if followed by success in coping with it – could produce an enhanced self-image and greater self-confidence, which, in turn, could promote health and longevity. This may be a factor of some importance in the longevity of those who attain a particularly advanced age. But, overall, the effect of stress on health seems unlikely to be other than negative.

Contrary to widely held expectations (see, e.g., Crimmins 1986, Brody 1987), the recent gains in life expectancy at the upper ages may prove to be a poor guide to the future. There is reason to believe that mortality among the elderly in low-mortality populations, rather than decreasing still further, will actually increase within another decade or two – even in the absence of war or geophysical catastrophe. In its pattern of mortality, it is by no means impossible that the current generation of elderly in these populations will prove to have been unique: unique with respect to the differences between them and those who will succeed them as old people, and unique with respect to the allocation of social resources.

In their mortality experience, those in low-mortality populations who are currently at the more advanced ages appear to be reaping the benefits of a particular combination of conditions and experiences both new in world history and short – perhaps no more than two or three generations – in duration. To begin with, these old people are biological survivors; survivors of higher levels of mortality in infancy, childhood, and early adulthood. Those who succeed them will also be biological survivors, but, in contrast, biological survivors of a period of particularly low mortality levels; of a period when unusual efforts were made to ensure the survival of weak and underweight newborn, of accident victims, of those with physiological and circulatory disorders. Although the 'insult accumulation' model would predict higher mortality among today's generation of elderly than among its successors, the alternative 'constant frailty' model, predicting lower mortality (Alter & J. C. Riley 1988), seems in the circumstances to be the more applicable.

Today's old people are also emotional survivors: of two world wars and at least one depression; in some cases, of civil war and revolution, ruinous inflation, repression, and political upheaval; of extensive urbanization and residential change; and of the rapid growth of material consumerism, with its inherent threat to traditional behavior and values, to cooperation and understanding between generations, and to satisfaction with group goals.

Compared to those who will succeed them, the present generation of older people has also had less exposure to certain of those deleterious environmental conditions the unhealthful consequences of which ordinarily manifest themselves only after many years of exposure: smoking, asbestos fibers, carbon monoxide, atmospheric lead, radioactivity, toxic agricultural materials, for example. They have also had arguably less exposure than will succeeding generations of elderly to certain types of emotional stress associated with family breakup and reconstitution, and with new role expectations and statuses for women, the consequences of which (if any) for longevity have still to be ascertained. And a higher proportion will have spent a larger part of their lives in rural and small-town environments characterized by higher life expectancies.

The present generation of elderly has also benefited from the allocation of social resources in ways likely to effect reductions in mortality at both the highest ages and the lowest. This has been possible, in part, because, compared to those who will succeed them, these aged have been relatively few in number and a relatively small proportion of the total. It has also been possible because, until recently, major gains in reducing infant and child mortality could still be had with relatively small outlays of money and effort. Neither of these conditions is likely to continue. The aged may gain something in political strength as a result of their increased proportions in the population, but any such gain is unlikely to be enough to produce the reallocation of resources (even if the aged wanted it – and it is by no means certain they would) necessary to procure the blessings of additional years of life to such a sizable number – a number that, because of its higher concentration in the oldest ages, may also, on average, be less fit. It is also unlikely to be enough to counteract the opposite pressure to allocate greater resources to the care and training of the relatively less numerous – but arguably more socially important – populations of children and youth.

Death will bring sorrow and bereavement, whatever the age at which it occurs. It may also bring economic hardship and usually, as well, the loss of certain, mostly personal, services, like companionship, intellectual stimulation, care and counsel. But if death can be postponed to at least the early 70s, as it now is for the majority in low-mortality populations (for approximately two out of three among men, and three out of four among women), it is unlikely to occasion either as much disruption or as much loss – whether to individuals or to the society as a whole – as when it occurs at younger ages. From the standpoint of both public policy and the more individual concern for one's fellow human beings, the most important questions about the elderly – as with any sector of the population – relate to: (a) their physical and mental condition, (b) their ability to cope with their own individual condition and the social setting in which they find themselves, and (c) the ability of the society, in turn, to cope with *them*. It is less mortality than morbidity and disability that raises the truly important questions. Irrespective of what happens with the fertility component of the equation, to the extent that declines in mortality rates at the uppermost ages level off or cease – especially if this occurs with no worsening of general levels of health – the degree of demographic 'deterioration' so feared in some quarters on the basis of present trends will be just that much further attenuated.

REFERENCES

Ahlburg, D. A. & J. W. Vaupel 1990. Alternative projections of the U.S. population. *Demography* 27(4), 639–52.

Alter, G. & J. C. Riley 1988. Frailty, sickness, and death: models of morbidity and mortality in historical populations. *Population Studies* 43(1), 25–45.

Benet, S. 1974. *The long-living people of the Caucasus*. New York: Holt, Rinehart & Winston.

Brent, L. 1986. Approaches to tolerance in man. *Nature* 321, June 12, 650–1.

Brody, J. A. 1987. The best of times/the worst of times: aging and dependency in the 21st century. In *Ethical dimensions of geriatric care*, S. F. Spicker *et al.* (eds.), 3–21. Dordrecht, Netherlands: D. Reidel.

Butler, R. N. 1975. *Why survive? Being old in America*. New York: Harper & Row.

Caselli, G. & J. Vallin 1990. Mortality and population ageing. *European Journal of Population* 6, 1–25.

Compton, P. A. 1985. Rising mortality in Hungary. *Population Studies* 39(1), 71–86.

Crimmins, E. M. 1986. The social impact of recent and prospective mortality declines among older Americans. *Sociology and Social Research* 70(3), 192–8.

Day, L. H. 1988. Numerical declines and older age structures in European populations: an alternative proposal. *Family Planning Perspectives* 20(3), 139–43.

Demeny, P. 1987. Pronatalist policies in low-fertility countries: patterns, performance, and prospects. In *Below-replacement fertility in industrial societies: causes, consequences, policies*, supplement to Vol. 12, *Population and Development Review*, K. Davis, M. S. Bernstam & R. Ricardo-Campbell (eds.), 335–58. New York: Population Council.

Ermisch, J. 1981. An emerging secular rise in the Western World's fertility? *Population and Development Review* 7(4), 677–84.

Gerth, H. & C. W. Mills 1954. *Character and social structure*. London: Routledge & Kegan Paul.

I.N.E.D. (Institut National d'Etudes Démographiques) 1989. Dix-huitième rapport sur la situation démographique de la France. *Population* 44(4–5), 711–76.

Kebric, R. B. 1988. Old age, the ancient military, and Alexander's army: positive examples for a graying America. *The Gerontologist* 28(3), 298–302.

Keyfitz, N. 1972. On future population. *Journal of the American Statistical Association* 67(338), 347–63.

Keyfitz, N. & W. Flieger 1968. *World population*. Chicago: University of Chicago Press.

Lederberg, J. 1988. Medical science, infectious disease, and the unity of mankind. *Journal of the American Medical Association* 260(5), 684–5.

Nossal, G. J. V. 1987. Current concepts: immunology: the basic components of the immune system. *New England Journal of Medicine* 316, 1320–5.

Potter, D. M. 1954. *People of plenty*. Chicago: University of Chicago Press.

Preston, S. H., N. Keyfitz & R. Schoen 1972. *Causes of death: life tables for national populations*. New York: Seminar Press.

Prominska, E. 1990. Demographic perspectives on stagnation and change in Polish society: comparative perspectives. Paper presented at conference held at Rockefeller Study and Conference Center, Bellagio, Italy, February.

Ryder, N. B. 1969. The emergence of a modern fertility pattern: United States, 1917–66. In *Fertility and family planning*, S. J. Behrman, Leslie Corsa, Jr. & R. Freedman (eds.). Ann Arbor: University of Michigan Press.

Ryder, N. B. 1975. Notes on stationary populations. *Population Index* 41(1), 3–27.

Taeuber, C. & I. B. Taeuber 1958. *The changing population of the United States*. New York: John Wiley.

U.N. (United Nations) 1949/50, 1957, 1959, 1960, 1965, 1983, 1984, 1985, 1986. *Demographic yearbook*. New York.

World Bank 1986. World population projections. Unpublished. Washington, DC.

W.H.O. (World Health Organization) 1980–. Environmental Health Criteria series. Geneva: Division of Environmental Health, World Health Organization.

3

THE CHALLENGE OF NUMERICAL DECLINE AND OLDER AGE STRUCTURE:

Part 1 Finances and the provision of care

INTRODUCTION

Whatever their magnitudes, the numerical declines and older age structures expected for these low-fertility/low-mortality populations will bring problems; problems of the inefficient use of capital equipment and lost economies of scale, for instance; of finances, of the provision of adequate services and care, of accommodating to changes in role relationships and in family and household composition. These problems are not beyond solution, nor must they necessarily be of the magnitude they commonly are – or are expected – to be. But they should not be minimized. A higher proportion at those ages most associated with physical and mental deterioration, bereavement, reductions in income and status, and frequently sudden and marked changes in living arrangements and social roles is hardly a joyful prospect. There will be compensations, of course, at least from the standpoint of society in general (about which more in Chapter 6); but, on the whole, it is a set of conditions likely to elicit calls less for celebration than for accommodation – assuming it does not first produce a paralysis of pessimism instead.

From the standpoint of their more general social significance, two of these problems – finances and the provision of personal care – are of particular concern. It is these that will be discussed here, keeping the matter in context by focusing as much as possible on the whole of the society and not simply on its more elderly members. Certain other problems (and issues) will be addressed in Chapters 4 and 7.

FINANCES

Introduction

In this age of government-supported social security programs and seemingly ubiquitious economic indicators, discussions of age structure quickly get on to problems of finance. The higher expenditures associated with old age relate not only to pensions and the like for which there is an age qualification, but also to the various programs and services that address the sorts of needs – for medical care, rehabilitation services, and nursing homes, for example – generally more prevalent among older people. Just how much the aged cost relative to other sectors of the society is the product of many things: of their numerical share of the population, certainly, but also of, among other things, the pension and social service systems applicable to older people and the efficiency with which these are administered, the attitudes and practices of younger sectors of the population concerning older people, the range of available lifestyles, and the relative physical and mental condition of the aged (itself a function of a whole host of conditions – from past and present decisions of the aged regarding lifestyle to the character of the social and natural environments in which old people find themselves). There is much more to these costs than simply numbers and age distributions.

Pension and social services systems have a particular significance in industrialized countries. Not only are these the countries best able to support such systems, they are also, because of their demographic structures and the fact that the social and economic changes they have experienced have attenuated alternative sources of support, the countries most in need of them. It has been claimed (Wilensky 1975, 47–9) that older age structures and the aged themselves are a major force for further development of these systems. It would be surprising if this were not so. Yet countries of similar age structure exhibit a considerable variety of such programs, and especially of rates of expenditure on them; and there have been marked changes – both expansions and contractions – in individual countries' programs over very short periods of time (U.S. Department of Health, Education, and Welfare, Social Security Administration, Office of Research and Statistics 1967 and 1975, L. H. Day 1978). Castles' (1982) emphasis on the importance of 'partisan control of government,' as against demographic change, in determining the nature of such programs and the level of their public support appears well placed.

Still, whatever the dimensions of these programs at any one time, the aged will on the whole be taking more from the public purse than they put in. The major charge is pensions – even though it is public expenditure on health and custodial services that looms particularly large in discussions of the costs associated with older age structures. If data for those of these countries that are members of O.E.C.D. can be taken as a guide, it is probable that in all of these countries (except Canada, where the amount spent on education is slightly higher), expenditure on income maintenance for the aged exceeds that for any other social program (education, family benefits, health, or unemployment). For the O.E.C.D. countries, the range in 1985 was from 26 per cent of all social expenditure in Canada to 49 per cent of it in Italy, with a median of 40 per cent (or, if expenditure on education is excluded, from respectively, 37 to 62 per cent, with a median of 52 per cent) (calculated from data in O.E.C.D. 1988a, Table B.1).

Nor do current demographic trends imply any early relief. Those post-World War II 'baby boomers' will enter their 70s in the second decade of the 21st century, their 80s in the third, and their 90s in the fourth. Tomorrow's swollen cohorts of old people are already in the wings. Quite apart from the question of pensions, what this means for other costs associated with an older age structure is that the age groups undergoing the most rapid rates of numerical increase will, for some decades into the future, be precisely those characterized by the highest per capita service needs (e.g., Manton & Soldo 1985, Rosenwaike 1985, Suzman & M. W. Riley 1985, Rabin & Stockton 1987, Rowland 1988).

Continuation of the recent tendency for the greater mortality declines to take place at the uppermost ages will only exacerbate the situation. But whether such continuation eventuates remains to be seen. Current differential mortality levels both between and within populations (e.g., M. W. Riley & Bond 1983, Pamuk 1985 and 1988, Stamler 1985, Manton 1987, Troyer 1988, Rowland 1991) point in that direction, and most observers believe that it will (e.g., Ogawa 1985, Brody 1987, Guralnick & Schneider 1987, Nathanson & Lopez 1987, Rabin & Stockton 1987), but, as already noted (Ch. 2, above), a case to the contrary can certainly be made.

Relative costs

Still, whatever the future trend of mortality, the current aging pattern in industrialized countries can hardly help but result – and for many years into the future – not only in increased demand for health services generally (e.g., Australia, Social Welfare Policy Secretariat 1984), but in increases in demand for expenditures for age pensions and services specifically geared to older people in particular (e.g. E.P.A.C. 1986).

With health expenditures, the higher the age grouping, the higher, on the whole, is the per capita outlay. The proportion receiving care increases with age, and so also does the intensity and duration of the care they receive. Per capita health expenditure in O.E.C.D. countries is, on average, about four times as high for those 65 years of age and over as it is for those under age 65; among the 'very old,' it is 'substantially higher' still (O.E.C.D. 1987, 88). A detailed analysis of 1978 health care spending in the U.S.A., for example, found the average expenditure on persons 65 and over to be two and a half times that on persons 19–64 years of age, and seven times that on persons under 19 years of age (Fisher 1980). Compared with those of younger age, the elderly were far more likely to go to hospital and, once there, to remain longer; and far more likely to visit a doctor and to receive more services per visit. Moreover, although less likely to visit a dentist, when they did visit one, the services they received were more likely to be particularly costly ones. And, of course, the elderly were far more likely to be the recipients of custodial care, whether inside or outside a nursing home. The behavior patterns of elderly Americans in these respects do not differ in relative terms from those of their counterparts in other industrialized countries (O.E.C.D. 1988a, 33–4). In Australia, for example, an estimated 23 per cent of the federal government's total 1984–5 expenditure on health was directed to the less than 4 per cent of the population aged 75 and over (E.P.A.C. 1988, 56).

Now it is possible that these estimates of the relative costs of health and custodial care among the aged are somewhat exaggerated. To the extent this is so, the additional economic drain from this source expected to accompany the shift to an older age structure would also be exaggerated. Because national health insurance coverage in the United States extends only to the elderly, we could expect older people's health costs in that country to be more a matter of public record, and for that reason to be the costs about which there is the greatest public awareness. Yet, even in countries with more comprehensive coverage, older people are still apt to appear the greater charge on the public purse. Partly, this is because the custodial care older people receive is more likely to be charged for – given their relative lack of the kind of family support for un-remunerated custodial care that is available to those of less advanced age (unremunerated care supplied by parents in the case of children and youth, and by spouses in the case of younger and middle-aged adults). Partly it is because calculations of the cost of health care seldom include certain less direct costs that are apt to feature more with the young and middle-aged than with the elderly.

Certainly any valid assessment of relative costs among different sectors of a population requires casting a wide net; one extending not only to the hospital, nursing home, and prescription drug costs that are the usual focus of such enquiry – and that bulk particularly large in the cost of services to the aged – but to outpatient services and even, perhaps, to some estimate of the monetary value of services rendered the sick and disabled by kin and volunteer workers, as well. It is, of course, possible that such coverage might reveal an even wider expenditure gap between the aged and other sectors of the population. But if the calculation were extended still further – as it reasonably could be in pursuit of a truly valid assessment of relative costs – to include some assessment of potential losses in manpower and (related to this) in tax receipts in consequence of disability and ill-health, the expenditure gap might be appreciably narrowed, for it is in these categories of cost that the younger and middle-aged would figure far more prominently than the elderly.

An especially good example of what is missed by the usual, more restricted approach is provided by motor vehicle crashes. In nearly every industrialized country, these are the major cause of death among both men and women at age 5–14 and 15–24; and they are, among men, either the leading or next-to-the-leading cause of death at age 25–44, as well. Their incidence is particularly high among young men – by a factor of two to five times that among young women (calculated from data in U.N. 1985, Table 33). Methods of cost-accounting vary, but the Australian government, for example (Australia, Bureau of Transport and Communications Economics 1989), estimates that the purely financial cost of road crashes, each year, is some $350 for every man, woman, and child in the country, with each road death costing some $450,000. This is costly enough. What is important for the present discussion is that there are other results than death. While the number of motor vehicle deaths in Australia is some 3,000, each year, the number of injuries from this source amounts to some 80,000 – of which 30,000 are serious enough to warrant hospitalization (*ibid.*). Among these are a significant number of debilitating head and spinal injuries. From the standpoint of age structure and finances, the important point about these injuries is that most of them involve not simply young males, but, in Australian experience, at least, young males who live about as long as do others their age who remain free of these afflictions (Yeo 1985, and Broe 1986, cited in Cass, Gibson & Tito 1988). Not only do fewer of the elderly experience such injuries, but when they do experience them it is ordinarily after, rather than before or during, their periods of employment; and, because of their shorter life expectancies, their experience of these disabilities entails, on average, fewer years of life characterized to some degree by dependency and the necessity of being cared for.

Origin of higher costs for the elderly

Nevertheless, exaggerated or not, the per capita health costs of the elderly exceed those of every other sector of the population. Nor do these costs occur at a stage of life when they might be defined as part of the cost either of preparing the recipients for economically productive lives or of enabling them to continue with such lives. Instead, they occur when the economically productive stage of life is ordinarily thought of as long past. From a purely economic perspective, these costs thus represent neither capitalization nor maintenance; only costs, pure and simple.

Yet they are not carved in stone. Reductions are surely possible. Nor is it out of the question to expect some mitigation of the undesirable consequences of the conditions in which these costs originate, not to mention a reduction in the incidence of the conditions themselves.

Three elements underlie these higher health expenditures for the aged: (1) higher incidences of ill-health and incapacity – even if there is necessarily considerable uncertainty surrounding the measurement of these incidences (Warren 1987), (2) higher rates of institutionalization and hospitalization, and (3) higher rates of death. Old people are more likely to suffer ill-health and incapacity; they are more likely to be institutionalized (whether because of illness or incapacity or simply for the sake of convenience) and, when hospitalized, to remain there longer; and they are more likely to die. Although these are, of course, causally related, it is possible for present purposes to address each in a manner in some measure exclusive of the others.

(1) Ill-health and incapacity. Like other organisms, human beings deteriorate with age. They lose teeth, hearing, sight, strength, stamina. They become prey to illnesses and conditions either seldom experienced by younger persons or experienced by them with lesser force. And they become more subject to mental confusion and brain failure, however lacking these elusive concepts might be in established definition and precise meaning (Carter 1981, 2–3).

But the extent of such phenomena varies widely among individuals. People do not age all in the same way. In this, as in so much else, the elderly are a notably heterogeneous lot (Rosenmayr 1981, M. W. Riley & Bond 1983 (cited in Manton & Soldo 1985), Manton & Soldo 1985, Rosenwaike 1985, Maddox 1987). Moreover, because their causes are only partly organic, and their manifestations perhaps even less so, ill-health and incapacity also vary widely in incidence between different societies and between social groupings within societies (see, e.g., Benet 1974).

Nevertheless, one thing is certain: while younger people may be

Table 3.1 Proportions with specified incapacities, by age and sex: Sweden, 1980–1

Sex	Age	% with reduced hearing[a]	% with reduced eyesight[b]	% with a serious mobility disability[c]
Males	16–44	3	–	1
	45–64	13	1	7
	65–74	19	2	19
	75–84	27	9	39
Females	16–44	2	–	1
	45–64	5	1	10
	65–74	11	3	21
	75–84	21	13	49

[a] Persons reporting that they cannot without difficulty (either with or without a hearing aid) follow a conversation between several other people.
[b] Persons reporting that they cannot without difficulty (either with or without glasses) read an ordinary newspaper.
[c] Persons reporting themselves as having to use, e.g., a stick, trestle or wheelchair, or as having to rely on another person, to move about within or without their homes.

more subject to the incapacities arising specifically out of motor vehicle crashes (as well as out of work experiences and sport), the overall prevalence of ill-health and incapacity is higher among the elderly. Moreover, whatever the social setting, and despite considerable variation among individuals, this prevalence among old people increases with advancing age (e.g., Brody, Brock & Williams 1987, Garber 1987, Guralnik, Brock & Brody 1987, Rabin & Stockton 1987, Australian Institute of Health 1988, ch. 4, Verbrugge 1988). Swedish data (Statistics Sweden 1982, available in U.N. Statistical Office 1989) illustrate the general pattern with respect to physical disability (Table 3.1).

A suggestion of the general pattern with respect to mental disability is seen in the following data from the early 1960s on the prevalence of moderate and severe brain failure or dementia among old people (both sexes combined) living at home in Newcastle, England (Kay 1980, cited in Carter 1981):

Table 3.2 Prevalence of severe brain failure or dementia among old people living at home: Newcastle, England, early 1960s

Age	% Prevalence
65–69	2
70–74	3
75–79	6
80+	22

American data illustrating the same association are available in Rabin and Stockton (1987, Ch. 4), and cross-cultural in Kay (1980).

The only exception to this association are certain psychiatric disorders that generally decline with age (Rabin & Stockton 1987). Yet, even here, senile dementia and Alzheimer's Disease, the two most prominently associated with old age, follow the general pattern. In fact, there is a suggestion that these two disorders, because they are associated with much higher rates of institutionalization, actually extend the lives of their elderly victims by protecting them from the normal stresses of life. Certainly the fact that they are not killers adds to the difficulty society has in adjusting to them.

In such matters, perception can introduce some confusion. Nationwide American studies based on self-reporting, for example, find that while few of the elderly are wholly free of the symptoms of at least one chronic disease (Rabin & Stockton 1987) – some related and some not related to death – few of them actually describe themselves as incapacitated or in ill-health. The explanation for this is not altogether obvious. Is it lower expectations associated with age or cohort? Is it having grown so accustomed to these conditions as to consider them normal and hence undeserving of mention? Is it that older people, defining themselves as particularly able to cope, are reluctant to admit to something that detracts from this image they have of themselves? If any of these, the future could see an increase in the demand for care in response to either higher levels of expectation or lesser acceptance of infirmity or ill-health.

Of the chronic diseases for which symptoms are reported, arthritis, which scarcely figures in mortality statistics, is the most prevalent and symptomatic. In fact, as a reason for medical care among older people, arthritis is outranked only by those two categories of major killer diseases: the cardiovascular and the neoplastic (Verbrugge 1988). From the standpoint of costs, this means that the trend toward an older age structure, especially when the highest rates of increase are occurring at the most advanced ages, is a trend toward higher proportions at those ages where the need for care is likely to be greatest (Manton & Liu 1984, cited in A. T. Day 1987).

(2)Hospitalization and custodial care. One consequence of the combination of physical and mental deterioration with (because of the death of the spouse and the departure of children) the disintegration of family units is an increase in institutionalization. The definitions vary somewhat, but in all the countries under consideration, the proportion 'institutionalized' steadily rises with age. In the U.S.A. in 1985, for example, the proportion resident in nursing homes (both sexes combined) rose from just over 1 per cent at age 65–74 to 6 per cent at age 75–84 and to 22 per cent at age 85 and over (N.C.H.S. 1987, 2).

The contribution to the generally more elevated health costs among the aged that comes from the costs associated with hospitalization exceeds that from the cost associated with custodial care. As already noted, old people are more likely to be hospitalized, and once in hospital, likely to remain there longer. The Australian data in Table 3.3 illustrate the general pattern (from Mathers & Harvey 1988, Tables 6.1 and 6.7).

Most of this higher rate of hospitalization can be attributed to the

Table 3.3 Hospitalization rates and average length of stay of elderly persons, expressed as a percentage of the rate at age 35–49: acute care hospitals, Australia, 1985

Age	Hospitalization rate		Average length of stay	
	Males	Females	Males	Females
35–49	100	100	100	100
50–64	180	105	142	140
65–74	310	146	198	230
75+	457	222	285	385

elderly's higher incidence of ill-health and general deterioration, as well as to their higher rates of death; some of it can be attributed to health practices specifically associated either with old people themselves or with the ills to which they are a particular prey; and some of it – how much varying substantially among different populations because of differences in both social policies and demographic composition – can be attributed to their household arrangements.

But hospitalization, regardless of who pays and whatever the patient's age or affliction, is expensive, especially in the more technologically advanced countries like those under consideration here. The thing to consider is, firstly, whether all this hospitalization is necessary and, secondly, whether what is necessary has to cost so much. Each of these points will be taken up in more detail later. For the present, it is enough to note that the answer in both instances is surely no, even if the complexity of the causal matrix precludes any precise accounting of extent.

(3) Death. Further to these phenomena of morbidity and incapacity are others more directly associated with death. In these societies characterized by the wide application of highly-developed medical technology, death, whatever the age at which it occurs, tends, like hospitalization, to be expensive. In part, this is precisely because death in such societies is so likely to occur in hospital (McCall 1984). Even if the costs of nursing home care and outpatient drugs are excluded, the costs of medical care are substantially higher among those who die than among those who do not (Scitovsky 1984). Nor, at least on the basis of American data, is this a recent phenomenon. Using data antedating the enactment of Medicare (a national program for the reimbursement of health care expenses incurred by persons 65 years of age and over), Timmer & Kovar (1971, cited in Scitovsky 1984) found not only that

(understandably enough) a substantially higher proportion among adult decedents than among the living population (73 per cent versus 13 per cent) received some care in hospitals and institutions over a 12-month period in the mid-1960s but that the median bill for such care was three times as high among the decedents as it was among those patients who did not die. Even among cancer patients, hospital payments were almost twice as high among those who died within the 24-month period 1969–70 than among those who did not (Scotto & Chiazze 1976, cited in Scitovsky 1984). In the late 1970s, among those enrolled in Medicare, Lubitz and Prihoda (1983) found that the average reimbursement for the cost of care was, among decedents relative to survivors, more than twice as high in the next-to-last year of life, and more than six times as high in the last year of life.

But at these societies' present mortality levels, only some 20 to 30 per cent among males and 11 to 18 per cent among females die before reaching age 65 (U.N. 1985, Table 36). This means that, with some 70–80 per cent of men and 82–89 per cent of women dying after age 65, the costs of death are particularly concentrated among those in the upper ages. The favorable mortality schedules giving rise to this phenomenon are something new in human history. They are, in fact, one of the greatest advances in human experience; and because their origin owes so much to human knowledge and effort, one of humanity's greatest achievements, as well. As recently as the beginning of this century, in Sweden, the lowest mortality country in the world at the time, the proportions of the total who would have survived to age 65 under the then current mortality schedule were only 46 per cent among males and 51 per cent among females; and still only 52 and 57 per cent, respectively, for those males and females who succeeded in surviving what were at the time very dangerous early months to reach their first birthdays (calculated from data in Keyfitz & Flieger 1968).

A prime element in the achievement of low mortality is the virtual elimination of death from those diseases (like measles, diphtheria, smallpox, scarlet fever, typhus, typhoid, tuberculosis, polio, pneumonia) that used to strike people down before they reached old age. The main killers in old age, accounting for some 80 per cent of all deaths at ages 65 and over, are heart disease, stroke, cancer, and diabetes: all diseases that, with the exception of certain forms of diabetes (by far, the least frequent killer of the four), either take a long time to manifest themselves or have an etiology requiring prolonged exposure to toxic materials. Toxic exposure is, in fact, particularly significant in the development of malignant disease, the cause of some 25 per cent of deaths among old people (or even more, if multiple causes are taken into consideration (Wrigley & Nam 1987)). So not only do nearly all the people in these low-mortality/low-fertility countries now die in old age, thus concentrating the costs of dying within a single age category, but when they do die, the cause of death is usually some disease that, by its very nature, is not one to which the young or middle-aged are particularly subject.

Narrowing the cost gap

What one perceives as possibilities for reducing the financial cost of the elderly is only partly a matter of data; it is a matter, also, of philosophical perspective, of ethics, of what is and is not to be included in the accounting of costs.

(a) Pensions. Although they are a prime element in the calculation of total financial costs incurred with an older age structure, pensions are in a uniquely different category of costs, and so will not be addressed in any detail here. It is not inconceivable that an adequate age-pension system could be supported solely out of savings during the period of one's employment (see, e.g., Vossen & Janssen 1987, Aaron, Bosworth & Burtless, 1989, Gonnot 1990). Such a program would require possibly a higher rate of saving during employment than occurs in most countries at the present time; and certainly either effective restraints on inflation and on the expansion of programs beyond levels sufficient for adequate – and equitable – provision for need, or else some mechanism capable of making such adjustments out of the sums paid prior to eligibility. If pensions are to be paid for out of savings rather than current revenues, benefits must somehow be balanced with contributions. Reforms undertaken by Japan in 1985, which maintain the existing level of benefit for the average insured person while reducing the future rate of increase in benefits (O.E.C.D. 1988b, 34) seem to suggest one way out of the difficulty. Another possibility is for payments into pension funds to be graduated upwards with age (Keyfitz 1989). If suitably related to ability to pay, this could reduce the relative burden on the younger worker while taking advantage of the fact that the older worker could be expected to have fewer expenses on behalf of childcare, and possibly housing as well.

Adequate pension systems funded solely out of savings during one's earning years exist at the private, individual level to at least some extent everywhere. They do not cover everyone, however, and they are not equitable. The private approach is necessarily limited. Where mortality is low, and the expected period of retirement correspondingly long, a private system depends on the potential beneficiary's having saved systematically at a fairly high rate over an extended period. This, in turn, depends on his or her having had a comparatively high rate of income and stability of employment (or at least of income) during the period when payments into the system were being made. It is a requirement most people cannot meet. No wonder, then, that truly adequate private pension programs are largely confined to the likes of academics, business executives, senior public servants, doctors and lawyers. For the remainder, hope must necessarily lie in national provision by governmental means. But at the present time this entails problems. One of these problems relates to the effects of inflation, which can widen the gap between the amounts paid into a fund during the period of one's employment and the amounts paid out of it during one's period of retirement. Another relates to the effects of peaks and troughs in age distribution combined with extensions of coverage. The marked increases expected in the numbers of old people at the beginning of the next century will come too soon for fully adequate savings to have accumulated in the pension funds of those countries whose programs are not of long duration. The problem in this instance lies less in the decline in numbers in the work force in consequence of low fertility than in the increase in the number of old people in consequence, partly, of declines in upper-age mortality, but, mostly, of the elevated fertility of the 'baby boom' period which extended from the mid-1940s to about the end of the 1960s. Expansion of these programs will add to their problems – whether that expansion represents an effort to remove deficiencies (through, for example, expanding coverage, extending the range of services, bringing payments into a more equitable relation to current wage and price levels), or whether it is but a response to greater political power of the aged (because there are more of them) or lesser political power of infants and children (that is, of their parents) because of low fertility. Whatever the motivating force behind it, and whatever its desirability, expansion will increase costs.

Of course, taxing pensions as part of general income – assuming a progressive tax structure – would retrieve part of their cost to the society. It would also foster greater equity. Claims that pensions are postponed wages or, in the case of universal age pensions, a reward for presumed contributions to the society are supportable enough, but hardly grounds for exempting them from taxation, especially when they may not be a pensioner's sole source of income. The further claim that taxing pensions, especially those privately funded by the recipient, is unfair because it penalizes savers in favor of spenders is less supportable. Beyond the question of whether the society actually benefits from the saving undergone (what kinds of activities, for instance, are fostered by the investments of pension institutions?), there is the question of equity: whether society as a whole should pay additional tax in order to reward still further those of its members who are lucky (or who receive better advice) in real estate or other investments, or who are better able to save because (whatever their individual merit or social contribution) they enjoy steadier, more remunerative employment.

(b) Reducing the incidence of ill-health and incapacity. Just how much of the present excess of ill-health and incapacity among the aged derives from old age itself (Manton & Soldo 1985), and how much from the life experiences and current living conditions of old people (Australian Institute of Health 1988, Ch. 5) is something of a moot point. Each plays its role. It is only the relative importance in particular instances that is subject to question. The data showing consistent increases with age attest to the importance of the former; those showing differences among populations (despite possible inconsistencies in definition and collection procedures), the importance of the latter. For present purposes, what is of interest is the possibility of reducing the incidence of ill-health and incapacity among the aged, irrespective of its origin.

The treatment and care associated with diseases and incapacities that are markedly more prevalent among old people tend to be costly. Because the very nature of these conditions largely restricts them to the elderly, many of the costs associated with them could be avoided if people would only die while still young – and do so quickly. Fortunately, there are less drastic alternatives – in curative health care, preventive health practices, and the introduction of certain changes into the social setting – capable not only of effecting some saving in costs but, also, of reducing the incidence of these conditions, or at least of mitigating the severity of the experiences associated with them.

Among curative health practices, intensive care units have achieved a remarkable degree of success with the victims of heart attacks, as have drug therapies in the treatment of hypertension and, in that way, in the treatment of heart disease and the prevention of stroke. Drug therapies have also done much to assist in the treatment and care of persons suffering from mental illness, and there is reason to hope they will eventually achieve some success with Alzheimer's Disease. Heart bypass surgery undergone by persons in middle age, while expensive, has undoubtedly reduced the ultimate costs (financial and emotional) of caring for these people in old age and, by enabling them to play a full range of social roles, further reduced the burden they might impose upon society. Laser procedures, already successful in the treatment of gall and kidney stones (and far less costly then bypasses) are now being developed to the same end. Relatively inexpensive cataract surgery for the eye has had similar results. The list is gratifyingly long.

But important as curative or ameliorative approaches might be (particularly in the individual case), it is prevention that offers the palpably greater rewards – in both cost reduction and better health. The preventive alternative will be discussed more fully in Chapter 7. Essentially, it involves changing the social setting in ways that (1) lower the incidence of ill-health and incapacity, (2) reduce aggravation of those conditions as already exist, and (3) assist persons with these conditions to remain independent and participating members of society. Specific approaches range widely, from the likes of halting ozone depletion and removing encouragements to smoking, to improving the ventilation of buildings and increasing the accessibility of public transport. As far as individual behavior is concerned, the measures to these ends among the elderly may be summed up as: low drug intake, good diet, and exercise – and this at all ages and not just at the more advanced, for in no small measure is one's condition in old age a function of the life one has led. As it happens, much of what would improve the situation of the aged would more than likely, by ridding the social environment of various elements that stimulate and reinforce negative lifestyles (LaLonde 1974), improve the life of younger persons as well. As will be developed more fully in Chapter 7, what benefits the elderly usually also benefits the rest of the population.

Cure and prevention will both result in higher proportions living longer. While this could add to the financial burden of health care and accommodation, it is not certain that it would. The point is whether these additional years of life will be characterized by health and fitness or ill-health and incapacity, and this is a matter of dispute. Some argue the existence of a 'natural' duration of human life (one of approximately 85 years has been proposed on the basis of experimental studies of cellular senescene (Fries 1980)), with gains in life expectancy resulting not from any extension of the human life span itself, but from the progressive elimination of 'premature death.' In this view, continued increases in life expectancy would ultimately produce a situation in which nearly everyone survived from birth to old age and then died within a relatively short period of time. Those arguing against this view do so on the grounds that as no mortality (other than that from 'external' causes – namely, suicide, homicide, and accidents) occurs in the absence of disease, the fact that recent declines in cause-specific mortality rates have been as great (and sometimes greater) for the very old as for younger age groups not only suggests the possibility of still further extensions of human life but runs counter to the view that human life is 'naturally' limited. Those arguing against the existence of a 'natural' limit also find support for their view in the fact that postponements to higher age have occurred in deaths from certain diseases of which the overall incidence has actually increased.

From the perspective of an essentially fixed span of life, the view one takes of the trend to an older age structure is apt to be fairly optimistic. The additional proportions surviving to the upper ages are expected to be characterized by generally favorable levels of health and fitness, and improvements in prevention and cure are expected to reduce (1) the average period of diminished vigor among the elderly, (2) the average duration of their experience of chronic disease, and (3) the length of time they will require medical and custodial care.

But from the perspective of an essentially unlimited life span (or at least a much longer one than at present), one's view of the trend to an older age structure is apt to be pessimistic. Here, additions to the numbers of elderly are more likely to be seen as the products of a technological extension of senescence, with, as a result, higher proportions of the elderly living ever more years of life in conditions of chronic morbidity and functional disability (Brody 1985, Guralnick & Schneider 1987, Guralnick, Brock & Brody 1987, Rabin & Stockton 1987, Ch. 4, Verbrugge 1988). Special impetus for pessimism comes from a longitudinal study in the American state of Massachusetts (Katz, Branch, Branson et al. 1983, cited in

51

Guralnick & Schneider 1987), which found that from about age 70, women had no advantage over men in the average number of years of 'active life expectancy' (defined as requiring no help in the performance of such activities of daily living as dressing, eating, bathing, and transferring from bed to chair). Given the consistently higher life expectancies of women, this suggests that any increase in the proportions surviving to old age will but result in more person-years of incapacity among the population as a whole. A more recent analysis of self-report data from national health surveys in Japan, the U.S.A., Britain, and Hungary (J. C. Riley 1990) adds to the pessimism with its finding that the amount of sickness time at each age reported by the interviewees moved in a direction opposite to the direction of change in the risk of death at each age. Although the risk of becoming ill moved in the same direction as that of dying (but not to the same extent), the *durations* of periods of illness moved in the opposite direction. Thus, when there was a decline in both the risk of death and the risk of becoming ill, the risk of *being* ill increased. It is, of course, possible that this finding owes more to changes in perception than in actual condition, but data like these do strongly suggest that declines in mortality – especially when mortality is already at what, from a historical standpoint, is a low level – are not necessarily an indication of any improvement in health; that they may, in fact, indicate quite the opposite.

At any given age level, there is now more variation in individual health than there once was. This is because of two things: (1) medical progress, which has succeeded in keeping more of the frail alive (Porterba & Summers 1987) and (2) the fact that changes at one health level – whether in the incidence of disease, the duration of episodes of disease, or case fatality rates – will have an effect on other health outcomes.

Certainly the balance of present evidence supports the more pessimistic view. Death rates at the more advanced ages are declining faster than those at the less advanced, as is illustrated in the French data presented in Tables 3.4 and 3.5, below (from data in I.N.E.D. 1988, Tables 31 and 37), thus adding to the numbers reaching these more advanced ages, and the additions to total years of life this has resulted in have not, so far, been equaled by similar additions to years of life free of ill-health and incapacity. For reasons discussed above (Chapter 2), this could change, but if and when it does, it is unlikely to do so in an immediately obvious manner. We can expect having to make provision for

Table 3.4 Percentage increase between 1950 and 1985 in the proportions surviving to specified ages, by sex: France

Age	Males	Females
20	7	5
60	15	15
65	20	20
70	27	28
75	40	44
80	62	74
85	106	124
90	177	201

Table 3.5 Proportions alive on January 1, 1985 who would not have been alive if 1950 mortality rates had remained in effect, by sex and specified age: France

Age	Males	Females
0–4	7	6
10–14	6	5
20–24	4	4
30–34	1	2
40–44	2	2
50–54	3	4
60–64	7	7
70–74	15	17
80–84	29	37
90–94	64	63

additional person-years of elderly incapacity for quite some time still.

(c) Changing the living environment. A third element in reducing the costs attending ill-health and incapacity is the creation of an environment appropriate to meeting the needs of those experiencing these conditions so that, as much as possible, they will be able to look after themselves and play an active participant role in society. This is gone into in some detail in Chapter 7. For present purposes, it will suffice to list a few examples: (a) technological improvements in home and community, many of which, in the community at least, have already been put in place in response to the United Nations Year of the Disabled; (b) extension and improvement of access to public transportation; (c) further development of various occasional care services (meals-on-wheels, housecleaning and repair services, assistance with keeping household accounts, and walk-in health and counseling services, for example); and (d) the creation of neighborhoods that are more to human scale, more geared to the satisfaction of human (as opposed to commercial or vehicular) needs. Many of these provisions would benefit more than just the aged: the families of the aged, of course, by enabling their elderly members to be less dependent upon them, but, more important, society as a whole.

(d) Revise methods of payments. A further approach to reducing the cost of health and custodial care for the aged would be to focus on the method of payment: revising it in ways that reduced the demand for such care and transferred costs to some agency other than the society. Ultimately, there are but three possible sources of finance for aged care: (i) the aged themselves, through either a fee-for-service arrangement or a private program purchased out of personal savings, (ii) persons (mainly kin and informal social networks) who give their support without remuneration – but doubtless with varying degrees of enthusiasm, and (iii) taxpayers, through government programs. All three can be resorted to, and usually are – but to varying degrees in different places and times, as well as at different stages in the life course.

The main objection to placing primary reliance on fee-for-service is that it results in some needs not being met. It discourages people from seeking help that might benefit them, and prevents people from receiving help they need. Nor is the system actuarially sound, for monitoring is necessarily limited to the procedures and services provided that part of the population paying for them. And it is not a particularly efficient system, either, for it encourages development of facilities and procedures in terms of potential profit rather than need and rewards the practitioner less for prevention than for therapy and maintenance. Though hardly a matter of simple cause and effect, it should come as no surprise that, among the countries under consideration, it is in the United States, the country in which the fee-for-service system is most entrenched, that one finds health care absorbing the highest proportion of gross national product (O.E.C.D. 1987). With some elements of care – certain types of plastic surgery or particular standards of accommodation, for instance – such a system works efficiently and fairly. But as far as the general needs of the aged for health and custodial care are concerned, the fee-for-service system is, at bottom, grossly inegalitarian and, for that reason, unjust and unethical. If, on occasion, it stimulates people to take more responsibility for themselves and their health, the gains in this quarter can hardly compensate for the losses in the others: the unmet needs, the pain and anxiety, the stress occasioned family and friends, the accumulation of personal debt to pay medical bills, the larger social and ethical loss entailed in the denial to some, purely on financial grounds, of certain fundamental privileges enjoyed by others more financially fortunate.

Making the aged pay during their years of employment also entails some problems, mostly concerned with collecting enough to pay for those either not in the labor force or with low earnings, as well as to permit later upward adjustments to cover inflation and any desirable expansion of services. But it would be an essentially equitable system – if payments were made not on the basis of the individual's likely future requirements but on the basis of his or her current ability to pay. It would be tantamount to taking money out of general revenue. Whether it was equitable would thus depend on the degree of equity in the general tax system: equity as to both the tax law itself and the way that law was administered.

(e) Eliminate services and restrict access. Another suggestion arising out of the concern over the financial costs entailed in an older age structure is that society's more elderly members be denied certain medical services and procedures, or at least that their access to them be restricted (e.g., Thurow 1984, Fuchs 1984, Callahan 1987, McGregor 1989). Certainly it can be questioned whether certain expensive procedures, like cardiopulmonary resuscitation, computerized axial tomography (i.e., C.T. scanning), occult blood testing, and neonatal intensive care for very small infants need to be continued on quite their present scale (irrespective of the patient's age) (Duggan 1989). Rightly observing that it is the quality and not the quantity (i.e., the length) of one's life that society should promote, Callahan goes so far as to suggest that old people accept as a social duty the necessity of forgoing some of the more extreme and costly of medical measures. Noting that good health does not guarantee a good life, nor a longer life necessarily a better life, he asks (p. 212), 'Have we [Americans] ... possibly reached a point of diminishing returns – not for some individuals afflicted with terrible diseases, but for our common well-being, happiness, and prosperity?' Although expressing doubt about the claims of some that health care for the elderly is the cause of more difficult times for young adults, he observes (of the U.S.A.) that

By investing enormous resources in health, we have made it less of a problem for most people; and yet we still spend large amounts of money to do even better. In the meantime, our school and transportation systems have deteriorated, our parks and recreational facilities have not kept pace with population growth and public usage, basic scientific research (save for biomedicine) has lagged, many of our cities continue to deteriorate, our production and economic strength have declined, and a generation of young people find it more difficult than ever to marry, raise children, and buy a home.

He concludes that 'the marginal improvement in overall health *statistics* [italics added] over the next few decades, at increasingly greater cost, is not at all likely to be a key to the nation's greatness, civil peace, prosperity, or national honor, or even the happiness of daily life.'

The means to the goal of more restricted access are several: restricting eligibility on grounds of age or (as with fee-for-service systems) income, limiting the period of time over which services can be received, reducing benefit levels or, at least, not expanding them to include improved technologies or therapies. So far as the countries under consideration here are concerned, this would probably be much like the fee-for-service system, because the levels of economic development in these countries would permit those with money to obtain pretty much what they wanted through private means. To the extent this happened, it would entail some inequity, although it can be argued that, as long as this occasioned no reduction in the quality of services available to the population as a whole, such a degree of flexibility would be permissible.

Already the American state of Oregon imposes specific cutoffs on free health care, and San Francisco is proposing to follow suit – under the direction, of course, of a panel that includes both physicians and nonphysicians and is guided by an ethicist (Higgins, cited in *Current Contents*, 1989). As it happens, restrictions of this kind are intrinsic to every health and custodial care program, whether or not specifically acknowledged. These programs could hardly operate otherwise. A comparison of American and British hospital expenditures in the period preceding the changes introduced by the Thatcher Government (Aaron & Schwartz 1984) attributes much of the markedly lower per capita figures in Britain to the fact that the rationing policy applied by British physicians, although still informal – and unstated – was more extensive than that applied by Americans, and that the British public were on the whole willing to accept the word of their physicians about the desirability in specific instances of the services and procedures under consideration. The authors note lower frequencies in Britain with particular respect to: coronary bypass, kidney dialysis, parenteral nutrition, scanning, X-ray examination, and the use of intensive care units. What is important about restrictions is not the fact that they exist, but, rather, their nature and extent, the purposes to be served by imposing them, and who and what bears their brunt.

(f) Reducing the cost of health and custodial care for the elderly. Prevention, therapy, and a more suitable environment will reduce the level of need, but they will hardly eliminate it. What, then, about the possibility of reducing the costs of those elements of health and custodial care that remain necessary? Must such care among the aged be as costly as it is, irrespective of the extent of morbidity and incapacity?

Anything that improves old people's health or their ability to look after themselves will reduce their need for institutionalization; while improvements in therapy and rehabilitation, not to mention administrative efficiency, will shorten the length of time they will have to spend in such facilities, once admitted. What is involved is a complex causal web of interconnecting components: (a) the physical and emotional condition of the individual, (b) the quality and efficiency of available therapies and rehabilitation measures, and (c) the social setting whence the individual patient enters the institution and to which he or she can be returned. Change in any one of these will affect the frequency and duration of institutionalization in the society, and, through that, its aggregate cost. This is true of any age group, but for present purposes, the important point is that the potential gains to be had are more substantial among the aged than among other sectors of the society because of their greater amount of ill-health and incapacity and the greater likelihood of their living alone.

There can be little doubt that substantial savings are possible, and that these can be effected without reducing quality, unduly restricting availability, or unduly burdening (financially or otherwise) either the aged or their kin and friends. On the basis of nothing more precise than the differing proportions of gross national product spent on health care, and some international comparisons of the frequency of certain surgical procedures (Aaron & Schwartz 1984; O.E.C.D. 1987, Chs. 1 and 2), it is obvious that the extent of these savings must vary considerably among countries. The general approaches to this end are three: (1) rely less on institutionalization and the application of costly procedures and therapies, (2) achieve greater efficiency in what institutionalization and therapy is resorted to, and (3) rely more on the aged themselves.

So far as the cost of, specifically, hospital care is concerned, the evidence in an American study that medical interns and residents, full-time faculty members in internal medicine, and ward clerks working on medical units tend 'drastically' to underestimate the costs of the tests and other procedures they order (Shulkin 1988) at least suggests that more knowledge would lead to more caution in prescription and, therefore, to lower costs. Another possibility is suggested by results of the United States Government's Health Interview Surveys showing substantial increases in physicians' incomes over the last three or four decades despite little change in the rate of individual visits to physicians (U.S., Health Interview Survey).

A somewhat different emphasis on physician costs comes from a study of fee-for-service utilization in the Canadian province of British Columbia (Barer *et al.* 1988), where such utilization accounts for some 95 per cent of the province's total medical care expenditure. Between 1974–5 and 1985–6, this increased at the whopping rate of 5.3 per cent a year. A third of this resulted from population increase, but less than 8 per cent of it resulted from changes in age structure. In fact, the changes in age structure that occurred (mainly a decline in the proportion of the aged who were 85 and over) were in a direction opposite to what would have led to increased rates of usage. The major element in the increase – accounting for nearly three-fifths of it – was increased per capita use, particularly at the extremes of age: 0–1 and 75 and over. The authors do not rule out increased morbidity (as a consequence of the application of more effective, but non-life-extending, new techniques and knowledge) as a possible factor, but they suggest that, with a 'dramatic' increase in the supply of physicians and essentially no change in hospital capacity during the period, the most probable cause of the increased expenditure was no less than a change in the way the health care system chose to treat the elderly. The fact that the greatest increase in use was in the specialist care category – especially for diagnostic services, together with medical specialists, raidologists, and pathologists – lends credence to such a view. The study strongly suggests not only that the treatment accorded the aged need not, on average, have been as specialized and as costly as it was, but that it need not even have been as extensive. Referring to other Canadian enquiries as well as their own, the authors conclude that a more aged population 'will not alone create a crisis for medical care.'

While Barer, Pulcins, Evans, Heitzman, Lomas and Anderson hold increased intervention by physicians particularly accountable for this rise in costs, it is worth noting that, in at least one study, and specifically with respect to myocardial infarction, the introduction of new techniques, while it increased per capita physician costs, nonetheless produced a reduction in overall costs (Sawitz et al. 1988).

Another Canadian study (Tilquin et al. 1980), which compared aged persons in a variety of institutional and noninstitutional settings, found that, while those within institutions were, as one would expect, more in need of day-to-day assistance, the proportion who could manage on their own for most things was, nevertheless, 'very high.' Marshall (1980, 202), commenting on these findings, writes that many of the aged

> are highly medicalized: they are very closely tied into a medical care system, and many of them are quite heavily drugged. This is another instance of sex and age conspiring, for women and the aged are major target groups for the drug industry, and the aging of the female population represents one growing market area for those companies who profit from illness. As the birth rate continues to be low, we might expect an increased focus on the aged as such a market for drugs, and the danger of medicalizing the life cycle and treating aging as a disease is very high.

Further insight into this matter is afforded by Aaron and Schwartz' comparison (1984) of per capita hospitalization costs in Britain and the United States. The authors note that, for the three preceding decades, the cost of hospital care in the United States had been rising at a rate exceeding that of either inflation or population increase. The principal force behind this, they find, was the rapid growth of medical technology: some of it of undoubted diagnostic and therapeutic benefit, but a growing proportion of it yielding – and at high cost–little or nothing of perceptible value. Without any particular losses in quality, the British were spending less than half as much per capita on hospitalization (adjusted for differences in wage and salary levels) as the Americans. They were able to do this, the study finds, because their system of medical care differed from the American in a variety of ways conducive to cost-saving. Most notably there was in the British system: (a) less scope for the profit motive, whether on the part of practitioners, hospital managements, pathologists, drug houses, or insurers, (b) less incentive to order tests and consultations for the purpose of avoiding malpractice suits, (c) a greater tendency, in the allocation of scarce resources, to apply criteria that are socially rather than individually oriented (e.g., whereas per capita health expenditures for children in the United States were 37 per cent as high as those for prime age adults, the comparable figure in Britain was 119 per cent; or, again: presumably in part because the cost of caring for those with hip disease is so much higher than the cost of caring for those with angina, the British performed relatively far more hip replacements and relatively far fewer coronary bypasses, although the costs of both procedures were about the same), (d) a reservoir of public goodwill for a national essentially egalitarian health service that had its origins in wartime adversity, (e) a health service organized within a parliamentary democracy characterized by party discipline and solidarity rather than, as in the United States, by politicians individually subject to the focused pressures of highly-organized interest groups, (f) a seemingly greater willingness to accept a family physician's word that no further treatment is warranted, (g) a lesser gap between average wages and the pay scales of physicians, (h) a lesser willingness to resort to aggressive care of the terminally ill. As the authors put it: 'Doctors in the United States realize that aggressive treatment of many terminally ill patients is often pointless, but they cannot do less because of pressure from the patient's family, the fear of malpractice suits, and the threat of intervention by the courts on behalf of the patient' (Aaron & Schwartz 1984, 124).

Further indication of possible savings comes from Chassin and Brook's research into American practices (1988). Clinical materials provided by 819 physicians and 227 hospitals were analyzed by panels of 'nationally know expert physicians' with regard to indications for three elective procedures – carotid endarterectomy, coronary angiography, and upper gastrointestinal endoscopy. The findings were of 'significant overuse' of each: about two-thirds of the carotid endarterectomies (about 80,000 of which are performed each year in the U.S.A., at an average cost of $13,000) were done for 'inappropriate or questionable reasons;' as were some 17 per cent of the coronary angiographies (over half a million of which are performed each year in the U.S.A.) and the same proportion of gastrointestinal endoscopies (about a million of which are done each year in the U.S.A.). Were such rates of 'inappropriate' use to apply to all procedures, the authors suggest that the medical bill in the U.S.A. could be trimmed by some $50 billion annually.

Institutionalization provides another instance of cost-significant differences in health and care practices among countries, in particular between Western Europe, on the one hand, and Australia, Canada, and the United States, on the other. The industrialized countries of Western Europe – already some 40 years further along in the trend to an older age structure – have tended to have smaller proportions of their old people in residential long-term care facilities, while at the same time providing more for them by way of publicly-financed community services. There are differences, of course: the percent-age of old people accommodated in institutions, for example, having ranged, in recent years, from a West European low of 4–5 per cent in West Germany to a high of 9–11 per cent in Sweden (Doty 1986, cited in A. T. Day 1987). So far as the nature of institutionalization is concerned, the likelihood is much greater in Western Europe than in North America or Australia that an institutionalized person will be in a facility that is nonmedical rather than medical in orientation. While, for example, the proportion of the elderly (in this instance, those aged 65 and over) in *medically*-oriented long-term care facilities was, in 1985, the same in Sweden as in the United States, the Swedish proportion in *nonmedically*-oriented facilities (like sheltered housing or subsidized special homes for the aged), that year, was five times higher. In the Netherlands, where the propor-tion in medically-oriented institutions was but two-thirds that in the United States, the proportion in nonmedically-oriented was almost seven times higher (Doty 1986, cited in A. T. Day 1987).

This greater emphasis on nonmedically-oriented institutionaliza-tion exists in Western European countries despite their generally higher proportions of elderly. Differences in age structure account for little of the differences in extent and type of institutionalization. As with so much else regarding the aged, explanation of cross-national variations requires looking beyond biomedical factors: to specifically relevant factors like social policy, of course (surely a prime determinant of such differences), but also to the more general forms of family life and social support (A. T. Day 1987). For example, institutionalization tends to be strongly associated with living alone and with never having been married – apparently

61

because such persons, in the event of poor health or a slowdown in physical capacity, have greater difficulty mobilizing the resources that would enable them to take care of themselves. Being more likely to live alone (because of the death of a spouse) or in an institution, older people are more subject to hospitalization in the event of any increased need for care. However frequent the contact an older person might enjoy with kin, if he or she is not actually living with a relative, the likelihood (so far, at least) is much less that she will have ready to hand someone on whom there exists a prior emotional claim for care. Hospitalization is an obvious solution, especially when it involves little or no direct financial cost to the patient or her family. There is more than age structure to the explanation of international differences in both institutionalization and the costs of elderly care.

What mainly underlies the higher health-care costs to be expected from the transition to an older age structure would appear, on the basis of such studies as these from the United States, Canada, and the United Kingdom, to be not so much age structure itself as the health care system's choosing to treat the elderly in a more intensive, more specialized, high-technology way. Many older people apparently receive more treatment than they require (or than is good for them), and what treatment they do require is often more specialized and expensive than it needs to be.

But there is one aspect of the costs of treatment that such studies do not address, and that is research. The higher-cost institutions in an American study of the costs of dying (Bloom and Kissick 1980), for example, were all teaching hospitals. Presumably, at least part of their higher costs were occasioned by the research they engaged in. One always hopes that research like this will result in better care, better health, higher-quality life for the population as a whole. But the extra costs it entails – financial, emotional, physiological – are, unfortunately, borne in the short run all but exclusively by those undergoing the treatment – as well as by their families and friends. Insurance coverage can spread some of these costs more widely: over the public as a whole in the case of social insurance, and over those who are insured in the case of private insurance. But insurance covers only finances. The charges upon emotions and physique remain undiminished – and concentrated on the afflicted few.

Certainly it is possible to reduce the costs of health and custodial care for the aged without sacrificing quality. Similar opportunity doubtless exists at other age levels, as well (see, e.g., Sagan 1987,

Barer *et al.* 1988). But when it comes to undertaking specific action to this end, one wants to be able to distinguish those procedures and therapies that are truly conducive to human wellbeing (or, when the outcome is uncertain, those that are at least grounded in a concern for human wellbeing) from those with no more claim on our attention than that they were undertaken out of ignorance or inefficiency or in the pursuit of idle experimentation or, worse still, of personal power or financial gain.

The goal in individual longevity is not mere years, it is years of high quality. The concern about increased charges for care arising out of an older age structure is understandable, even if the likely costs are puny alongside those in certain other categories of public expenditure, such as military preparedness and highway construction.

(g) Reducing the cost of dying. Although there are substantial numbers at all ages – those experiencing a 'natural' death – whose dying affects the system very little, dying appears to have a greater effect on the utilization and cost of services than does aging (Roos, Montgomery & Roos 1987). It is sometimes forgotten that death can only be postponed, never prevented. As already noted, the marked postponement of death that has taken place in the countries under consideration is one of the greatest of human advances and achievements. And there are those who argue – on biological grounds alone – that still further postponement of death in these countries is possible (e.g., Schatzkin 1980), even likely.

Whatever the likelihood of further postponement, the point is that, the more it is postponed, the more is death likely to be caused (wholly or in combination with other causes) by something to which younger persons are not generally subject: causes that take longer to manifest themselves, and causes that require longer exposure to toxic materials.

A reasonably quick and painless death following upon a long and satisfying life must be something of a universal human goal: not so quick a death, however, as to preclude the possibility of the survivors (the family and friends of the deceased) making an adequate adjustment to their loss, or of the deceased putting his or her affairs into reasonable order beforehand – arranging for the disposal of property, making gifts and bequests, re-establishing good relations, visiting special persons or places, even getting on good terms with God.

As it happens – because of the development of medical technology and professionalism – a quick death brings with it an economic reward in addition to the psychic: the quicker and less painful it is, the less likely is it to be financially costly. That the causes of death of particular prominence among older people so often entail extended periods of ill-health and incapacity is thus doubly unfortunate: financially as well as physiologically and emotionally. Nevertheless, death at any age – especially in medically developed societies – tends to be expensive (Lubitz & Prihoda 1983; Lamm 1984). Postponing it to old age but heightens the seeming disparity in financial burden between the elderly and the rest of the population. Quite apart from whether anyone would have it otherwise (the alternative being death at an earlier age), is death at older ages more expensive, *per se*? On a cause-specific basis, that is, does death cost more when it occurs to an older than to a younger person?

It is conceivable that there could be age-related differences in the extent to which physicians attending a dying patient might avail themselves of the opportunity thus afforded for experimentation and research. But what direction these differences might take (assuming they even exist) – whether, that is, toward increasing costs more among older or among younger patients – is open to question. A more general consideration is that it is not always possible till after the fact to know that it was, indeed, impossible to postpone death to a degree – and (one would hope) at a level of personal functioning and good spirits – sufficient to compensate for the expenditure entailed in time and money and in the lengthened period of stress and anxiety for the individual and his or her family and friends. In cases of such uncertainty, it seems reasonable to suppose that any greater expenditure of time and effort would more often accrue to the younger than to the older patient, if only because the younger will be seen as having prospectively more years of life to gain.

Such conjectures are not altogether beside the point, but they necessarily relate to only a small proportion of deaths. In Australia, in 1986, for example, half of the deaths to persons 0–44 years of age were attributed to the same general categories of cause as those that accounted for nine out of ten deaths among persons 65 years of age and over; yet while these deaths among those 65 and over were almost two-thirds of all deaths, that year, the deaths attributed to these causes among persons 0–44 years of age were less than 5 per cent of the total (calculated from data in Australian Bureau of Statistics 1988, Table 3). In and of itself, age seems unlikely to have much bearing on the cost of dying.

Irrespective of cause and timing, must the inevitable – death – be as expensive as it now is? Not always knowing beforehand whether it will occur (Scitovsky 1984, Levinsky 1984) presents problems of knowing what can – or should – be done to effect cost savings on services and procedures in specific instances. Nevertheless, experience is accumulating to support the view that, if at all possible, it is less expensive (and often emotionally preferable, as well) to die at home or in a hospice instead of a hospital.

An American study is instructive on these matters. In a comparison of the costs of care at home and in hospital in the last two weeks of life for 12 matched pairs of males and 15 matched pairs of females (median age: 68; age range: 35–94) terminally ill with malignant disease (Bloom & Kissick 1980), hospital costs were found to exceed home costs by a factor of 10.5. And what did this additional expenditure buy for the hospitalized patients? A 'remarkable' quantity and breadth of services in the form of more intensive technologic measures – diagnostic, therapeutic and palliative. It also bought – and this borders on the macabre – the administration of diagnostic and therapeutic services to nearly all of the hospitalized patients 'until the day of death,' despite the fact that, according to the physicians working on the study, during the last two weeks of life, most physicians would know that death for such patients was 'imminent.' In most cost categories – physician services, diagnostic and therapeutic radiography, physical therapy, respiratory therapy, blood transfusions, and medical supplies, for example – the costs incurred by the hospitalized patients were 'substantially larger.' By malignant site, the cost differential ranged from being 4 to being almost 74 times as high for the hospitalized patients. As for diagnostic procedures (tests and the like), while the cost of these averaged less than 1 per cent of all charges among the home patients, it was nearly 19 per cent among the hospitalized. For all forms of palliative therapy the cost averaged 12 per cent of total charges for the home patients as against 29 per cent for the hospital patients. The difference cannot be explained by insurance coverage, for all the patients in the study had access to a comprehensive Blue Cross or Medicare home care program involving no additional out-of-pocket expense. On the other hand, self-selection, so far as carers are concerned, obviously was a factor: all the survivors of the home care patients reported that providing care for their family members had been a 'satisfactory experience.' One item of cost that might have been higher with the home patients was the loss of income on the part of family members who stayed away from a job in order to administer care, but this was not enquired into.

If, in this age of high technology, alternatives to dying in hospital are to become more prominent without jeopardizing quality of care, care and therapy will have to be administered within well-developed ethical guidelines. These should, as a minimum, be predicated on respect for the dignity of the individual and on the goal of minimizing pain and suffering on the part of both the patient and his or her family and friends. This would be desirable irrespective of the consequences for costs. That it would probably result in lower costs could spur the undertaking.

Leaving aside the question of attaining a satisfying life, how does one ensure a quick, painless death? How, that is, does one, at a minimum, ensure, firstly, that one's death is not preceded by an extended period of disability or pain and, secondly, that one dies (that is, that one is allowed to die) in one's own time and with dignity. In attaining the first, the major means are prevention and therapy; in attaining the second, they are euthanasia and suicide. Euthanasia, which embodies both active and passive approaches (Lamm 1984) is, of course, a matter of degree, ranging particularly widely nowadays as a consequence of modern technologies and the increased role in caregiving played by professional practitioners. Suicide is also a matter of degree (L. H. Day 1984), but to a far lesser extent. If we are to be in any way guided by Seneca's eminently reasonable and practical observation that 'the wise man will live as long as he ought, not as long as he can' (quoted in Hook 1987), we will need the assistance of both prevention and therapy, on the one hand, and a reasonable degree of access to euthanasia and suicide, on the other. Voluntary euthanasia is already practiced 'quite openly' in the Netherlands (Kuhse 1987, cited in Kuhse & Singer 1988), and recent research in the Australian state of Victoria suggests that medical practice is not only moving in that direction there, as well (Kuhse & Singer 1988), but that it would proceed even further along this route if the law were changed. It remains to make those changes in law and custom that will enable individuals to escape voluntarily from unrelievable pain and suffering, either with or without the assistance of others (see, e.g., Bates 1980, Hook, 1987).

General considerations

It is sometimes argued (e.g., Davis 1988) that the amounts spent on pensions, health and custodial care for the aged are essentially transfers from a society's economically productive to its economically unproductive members. But these sums can also be seen as deferred wages (in those cases where the recipient has been in the labor force) or as a reward for a life of service to society: services in the form of raising the next generation, maintaining community and social cohesion and continuity, performing myriad types of acts of benefit to society, performing the many roles that define a society and sustain it.

Nor is it as though the aged have not paid for what they get through a lifetime of paying taxes and insurance premiums, a lifetime of work and effort – whether in or out of the labor force – of benefit to society; or that they are not still contributing to society in ways of economic as well as noneconomic significance (e.g., Zapf 1984, Wu 1986): caring for themselves and one another (the aged are, in fact, their own chief source of carers), caring for others in ways extending from childcare to the performance of a variety of community services, or extending financial assistance to kin in the form of, for example, down payments on mortgages, educational expenses for grandchildren, and assistance in emergencies (A. T. Day 1985). Intergenerational wealth flows – whether economic, social, or emotional – are by no means all in one direction. That so few of these contributions of the aged are distributed through the marketplace is no sign that they lack value or significance.

Surely, one mark of a healthy society is that its members contribute to the wellbeing of others than themselves, and, moreover, that these contributions extend across generations. As it happens, many of the additional costs associated with a more aged population structure have their origin in the welfare state; in, that is, the use of deliberate political action to distribute more equitably what has already been distributed (presumably less equitably) through the market (Ringen 1987, 2–5). It is a development most people probably approve of. Yet, there is in so much of the discussion of the costs associated with older people the hint of a kind of upside-down egalitarianism. Rather than anything particularly special, so the reasoning seems to go, old people are much like everyone else – only uglier, less fit, more forgetful, less productive, more expensive. And their needs, far from normal and appropriate to a stage in life to which most people in low-mortality populations can nowadays reasonably aspire, are seen as somehow abnormal and in competition for their satisfaction with the needs for other (more attractive, more productive) sectors of the society. The very term, 'dependency ratio,' so commonly applied to the ratio of the number of old people to the number of younger people in the society, implies not only a flow of goods and services in only one direction but, also, a body of recipients of these goods and services who not only are not altogether deserving of them, but also not doing much for themselves, either.

Concern over the costs of programs for the elderly is, however, only partly a matter of economics. It is a matter, also, of the relative novelty of some of the programs and the newly-arisen magnitude of the demand for them; of developments in medical technology and the higher levels of aspiration they lead to; and of general economic conditions, not the least of which is inflation. Something can be said, also, about perspective. We might well allay much of this concern if we were able to express costs not in dollars but in, say, 'missile-' or 'bomber-equivalents,' instead.

Moreover, it might be perceptually useful to cast a wider net when assessing costs: going, for example, beyond the hospital and prescription drug costs that are the usual focus of enquiries into health expenditure to include outpatient services, nursing home care, and even, perhaps, some estimate of the monetary value of services rendered by kin and neighbors in caring for the sick and disabled at all periods of life and not just old age. On a case basis, disability is apt to be more costly at younger than at older ages because of the twin possibilities that it will both last longer and have a greater effect on participation in the labor force. That modern welfare policies cover the major health costs of the elderly more often than they do those of other sectors of the population can exaggerate the apparent economic drain to the society inherent in a relatively older age structure.

How much older people, relative to other age groups, actually receive from society will be a product of their relative political strength, on the one hand, and the society's priority structure, on the other. At a less general, more specific level, the differences, so far as monetary costs are concerned, will relate to differences in patterns of employment, wage levels, and the experience of inflation. Overarching the whole, setting the range of limits for these differences and reflecting the society's priority structure, is policy.

While it is conceivable that the values of monetary reserves collected for age pensions and health benefits could keep pace with inflation, it is unlikely that they actually would. Whatever the method of accumulation, in most cases inflation will mean that the appreciation in value that can be expected from sums paid in at an earlier time will have to be supplemented to some extent from current payments if they are to have the same *relative* purchasing power as when originally collected. Similarly, if during the years when they were paying taxes and insurance premiums on behalf of future age benefits, the elderly received comparatively low wages, or experienced comparatively long or frequent periods of unemployment, any current program of benefits – if it is to be adequate – would require some compensatory supplement from current collections. To the extent that current payers are themselves subjected to low wage rates or unemployment, the differential rate of return between older and younger would be even greater.

But the sudden increase (or expected increase) in the financial 'burden' of the aged in many countries owes less to the increasing numbers of old people or the experience of inflation or unfavorable employment conditions than it does to changes in policy – particularly the removal of what most people would view as deficiencies in earlier policies. These are of two main types, both conceivably one-off phenomena: (1) the institution or extension of pension and medical assistance programs for the aged without having first accumulated the major portion (or, sometimes, any) of the necessary funding for such programs through payments on the part of the beneficiaries during their working years, and (2) the indexing of pensions to price or wage rates at a time of inflation – again, without any prior payments from the beneficiaries sufficient to support such an adjustment. Because inflation emphasizes the need for it, the incentive to indexation comes at precisely the time when it is most costly to implement.

Then there are those costs that arise from deficiencies in these societies themselves. These will be addressed in some detail in Chapter 7. For now, it is enough to note that, irrespective of their proportion in the population, the degree to which the aged are a social burden is, essentially, a function of the society's success in ensuring the wellbeing of all its members, and not just those of them in the more advanced ages.

Do finances really matter?

But having said all this about the cost of care, it remains to ask whether finances are, in the last analysis, really all that important. Is a lack of money really likely to prove a barrier to the provision of adequate care for the substantially older populations expected in the countries under consideration? There are two approaches to this question: one focusing on finances, the other on the nature of care.

Some (e.g., Crown 1984, Gibson 1989) argue against being concerned about the financial costs of aged care on the grounds that, whatever their level, such costs are unlikely to exceed what can be met by continued economic growth. This may have some potential validity, at least in the short run. But it is not a particularly reassuring view. For one thing, it rests essentially on a reification of the concept of economic growth – disregarding both the composition of that growth and the essentially arbitrary way in which it is defined and its rate calculated. For another, to rely on economic growth to ensure the provision of aged support is to rely on something that might not be altogether beneficial. Growth is not the same thing as development. As Herman Daly has pointed out, growth involves increasing physical scale: consuming more resources, producing more waste. Development, in contrast, involves achieving a qualitative improvement. So far, there appear to be no limits to qualitative improvement, but if there are, approaching or exceeding them can hardly entail any worsening of conditions. Approaching or exceeding the limits to growth, on the other hand – which are many, and becoming increasingly obvious – inevitably will (Daly 1986).

But it is questionable whether financing aged care (in the countries under consideration) need depend on economic growth at all. Given the comparative ease with which immense public sums can be found for military procurement or for superfluous and pretentious road systems (or, in the United States, recently, for repaying a massive debt arising out of highly questionable, if not outright dishonest, practices on the part of the executives of savings and loan associations), the prime determinant in the allocation of public expenditure would appear to be not so much the supply of money itself as the ordering of *priorities*; priorities, that is, among, if not the members of society as a whole, at least those with the power to set these priorities and act upon them.

What old people need most in the way of health care and custodial services is less the application of costly state-of-the-art techniques and approaches than the commitment of health practitioners and caregivers who: (a) are readily available when needed, (b) will treat their aged charges kindly and sympathetically as dignified adults, (c) will not begrudge these charges their time, and (d) will allow them to die when the time comes. This is not a description of something likely to be prohibitively costly in any financial sense. In terms of finances, about all such a regime would involve is getting away from, firstly, that tyranny of the technological imperative that argues that if it is possible, it must be done and, secondly, that tyranny of extreme egalitarianism that argues that if it is done for one, it must be done for all. The economic assessment of costs and benefits can be a useful tool in such undertakings, but the final criterion must be the general welfare of the society, for it is society that bears the ultimate burden (whether in terms of finances, emotions, or the quality of life). As in any instance of such decision-making, the problems are two: (1) who is to decide? and (2) what are to be the specific grounds on which decisions are made? (e.g., Levinsky 1984, Thurow 1984, Stone 1985). Appropriate ethical guidelines must be applied, of course; but only those of a very general character. Narrow, rigidly legalistic rules to be applied in carefully specified instances will not do.

THE PROVISION OF PERSONAL CARE

While the provision of appropriate care to the elderly is unlikely to be much limited by any genuine shortage of finances, the same cannot be said of personnel. A shortage of suitable personnel is likely to be of considerable importance. The problem can be solved, but solving it will depend essentially on changing the social and institutional setting so as to: (a) enable the aged largely to take care of themselves and (b) develop appropriate personality traits among both the aged and their potential carers.

While demographic conditions need not necessarily be prime determinants of the quality of people's lives, there can be no doubt that they establish the limits within which that quality is worked out. So far as personal care of the elderly is concerned, the significance of an older age structure lies in its implications for: (a) the availability of people to provide this care, (b) the availability of economic support for this care, and (c) the suitability of the physical and social infrastructure for meeting the needs of the elderly – both the needs met through the intervention of others and those met through the actions of the elderly themselves.

Having a higher proportion of old people in the population means not only a higher proportion, overall, who are likely to require some kind of attention, but a higher proportion, also, of non-aged people who will have a close relative experiencing such need. Shifting the age distribution toward the uppermost ages means that a higher proportion of the elderly will need some kind of assistance, and that this need will, on average, extend over a longer period of time.

Generally speaking, older people in industrialized societies seem to place great store on maintaining control over their decisions and activities. They see maintaining a separate household both as the norm for people in their age group and as the arrangement they most prefer for themselves (A. T. Day 1985a and 1985b). Moving in with children or other relatives is seen as something of a last resort (A. T. Day 1985b, Ringen 1987, 129–35, citing studies undertaken by the University of Stockholm School of Social Work). Although strongly desirous of remaining in close touch with family – especially with adult sons and daughters, as advancing years take their toll of spouses, siblings, and valued friends – most old people, when asked, express a preference for staying at home with outside help if they are unable to manage on their own. The goal is 'intimacy, but at a distance' (Rosenmayr & Kockeis 1968, Cherlin & Furstenberg 1986). By and large, a shared household usually reflects both the need for more comprehensive care and the existence of fewer financial resources to pay for services within the home. It might be noted in passing that, contrary to much economic theorizing, few of the aged – at least in the United States – save for the purpose of leaving an estate. Instead, the main reason they give for saving is to ensure their independence: the ability to live on their own, debt-free and capable of participating in the life of the community (Torrey 1988).

Yet, a variety of studies show that family members provide most of what help the aged receive with health care and daily activities, even those of the aged who are seriously ill or impaired (A. T. Day 1985b). Kin are at any age a major source of individual support. But among those in their late 70s and 80s, blood relatives, particularly members of the nuclear family, appear to be the primary source of such support (see, e.g., Litwak 1985). The number of brothers and sisters, for example, influences both the types of shared households potentially available in later life and the potential availability of nieces and nephews as a source of support, particularly if one is childless oneself.

But at the most immediate level, the major determinant of the amount and type of potentially available social support is the composition of the older person's household. Whether an older person lives alone, with a spouse, or with an adult child both reflects and conditions a variety of economic and social factors having important bearing on the nature of caregiving relationships. Living with a spouse, for example, generally maximizes the opportunity for receipt of substantial long-term personal care. Living alone, in contrast, requires a minimum capacity for self-help. Living alone also means that any assistance has to come from outside the household and, therefore, that there is the risk of intermittent supervision, of delays in the receipt of needed help, of inconvenience to the caregiver(s). As for living with an adult child, while this permits close supervision and immediate response to need, it also involves intergenerational relations at close quarters, which raises the possibility of stress arising out of the older person's dependency, as well as out of the role reversal likely to take place as the parent surrenders authority and decision-making and the child takes these on. For the primary caregiver, there is in the parent-with-adult-child household the further possibility of competition between the loyalties and demands of the child's spouse and children on the one hand and those of the older parent on the other. Recent social changes are likely to raise further problems. Particularly with respect to female kin (mainly wives and daughters) – who are the main, although by no means the only, source of household care of the aged – can such social changes as declines in fertility, increases in divorce and family instability, and the extension of women's labor force participation be expected to affect the willingness of potential caregivers to devote the time and effort necessary to sustain frail older relatives at home.

There is necessarily some uncertainty about specific situations, but, given the present significance of kin in the provision of aged care, a suggestion of what the implications of low fertility might be for the continuation of present arrangements can be ascertained by projections of future frequencies of kin under given fertility/ mortality assumptions. On the basis of continued low-mortality and 1978 American fertility patterns (which were, that year, at exact replacement level), Pullum (1982) has calculated that, eventually, about 11 per cent of the population would have no children, 7 per cent no siblings, 13 per cent no grandchildren, and 12 per cent no nieces or nephews. They would not be altogether bereft of kin, however: some 98 per cent would have at least one first cousin (with nearly a quarter having 10 or more), and hardly anyone would lack an aunt or uncle.

From the standpoint of providing care for the aged, the greatest importance in such projections presumably attaches to the proportions lacking both children and siblings. Greater concentration of fertility at 0–1 parity than in the base population used by Pullum (and this is currently the case in many of the populations under consideration here) would simply restrict the number of kin still further. Whereas, for example, Pullum's model was constructed on a population in which 11 per cent remained childless and some 7 per cent had no siblings, the comparable population in Hungary, although having essentially the same proportion (10 per cent) childless, would have twice the proportion (15 per cent) with no siblings (calculated from data in United Nations 1981, Table 50).

What such calculations emphasize is the need to determine the extent to which functions traditionally thought to be performed by close kin (spouses, children, possibly siblings) have been – or can in the future be – performed, instead, by more distant kin or even non-kin. As Wolf (1986) concludes from his projections of the lowest fertility levels in recent Dutch experience, which show that low fertility results in so many of the aged having no working-age family members whom they might turn to for support (he specifically excludes consideration of spouses), 'demographic realities preclude the use of the family as the sole source of support for the elderly.' Just how important – and necessary – then, is kinship to the provision of care to the aged?

It is widely believed that the social support of primary groups (among whom close kin figure prominently) is generally associated with better health and lower mortality. In effecting this end, such groups are seen as providing: (a) instrumental help, (b) information, (c) advice, and (d) emotional bonding capable of lessening stress or directly affecting physiological functions, like blood pressure and the immune system, that bear a particular significance to health and mortality (Berkman 1985, cited in Litwak & Messeri 1989). Yet, formal organizations are capable of providing the first three of these, and even, in some instances, the fourth – as when a formal organization reduces ill-health arising out of stress associated with unemployment by providing a job, or when a physician prescribes medication that controls blood pressure or improves the functioning of the immune system. Primary groups appear best suited to the performance of those tasks in which little technical knowledge is needed (Litwak & Messeri 1989 use the example of removing a lighted cigarette from the fingers of a sleeper) and there is such a high degree of uncertainty that an expert cannot be brought in in time to make a difference, while a person with everyday knowledge who is available can make at least some difference. Primary groups are seen as being more effective when managing nontechnical tasks because they are: less expensive (their ministrations do not entail the training and recruiting costs associated with more formal providers of care), less slowed down by the extended lines of communication associated with the detailed division of labor characteristic of formal organizations, and likely to be more conscientious because of being motivated by personal rather than economic ties (Litwak & Messeri 1989).

Quite obviously, the ideal system for aged care would combine the strengths of both primary and formal provision. While by no means unknown in the provision of aged care, the importance in this regard of strong primary ties to friends and neighbors is ordinarily secondary to that of close kin. Does this mean, then, that the absence of close kin must necessarily result in unmet needs? If it does, European populations are going to be in dire straits, indeed. They are, in fact, already well on their way, as the projections by both Pullum and Wolf amply show. Even in Australia, where the situation would be less extreme because of the, until recently, more elevated level of fertility, Rowland (1984) has calculated that, by 1976, a quarter of the women 70+ years of age lacked the potential support of sons and daughters, either because they had never married (11 per cent), had no children (11 per cent), or had (or would have) no children surviving to age 60 (3 per cent).

In addressing this question of the necessary significance of kin to the provision of aged care we must remember that existence does not always signify availability. Having even a close relative is no guarantee that help will be forthcoming from that person in the amount and of the type needed. In a study of elderly American women, Wolf (1988) found, for example, that, while seven out of ten of his sample had some problem with one or another 'activity of daily living' (that is, with feeding, dressing, bathing, getting out of bed, or going to the bathroom by themselves), fewer than half (44 per cent) of the children serving as these women's primary caregivers were actually providing this type of assistance. A study of the preferences of persons 45–65 years of age (the 'elderly of the future') in the Dutch province of Tilburg, Bartlema (1987) found two out of five expressing a preference for receiving help only from secondary sources, another two out of five for receiving it from both secondary and primary sources, and only one out of five for receiving it solely from primary sources. Quite apart from the possibility of being separated geographically or emotionally, one's kin may possess neither the living arrangements nor the money, time, physical or emotional stamina necessary to the task. It is not at all unusual, nowadays, for the child whom tradition might expect to shoulder major responsibility for the care of an aged parent to be a sexagenarian herself (and it usually *is* a woman) in need of care, or more than a little occupied with the care, say, of a disabled spouse, or of a grandchild or two so her divorced daughter can go out to work (A. T. Day 1985b).

Moreover, needs can be – and often are – met without the intervention of kin. Recent research on a nationwide sample of married and once married white women in the U.S.A. who were born in 1900–10 (A. T. Day 1991) found, for example, that almost a third of the noninstitutionalized who had a husband or a son or daughter named someone *other* than that person as the person 'most responsible' for providing them with 'practical help and support.' Some in this category had sons but no daughters, and were reluctant to turn to a son and daughter-in-law for help where they might have felt freer to approach a daughter and son-in-law. Others lived a half to two hours away from the nearest adult child. One out of seven of them had husbands who were ill or who for some other reason were considered by the respondents to be unable to take responsibility for support. In still other cases, the son or daughter who might have provided support was ill. And in a few cases, the respondent said she did not get along with either the adult child or the son- or daughter-in-law. Whatever the reason, the husband or adult child was not considered by these respondents to be an appropriate source of support.

Given present demographic trends in European populations, such findings have to be considered reassuring. While there is no denying that kin occupy a position of special significance in the provision of care to the aged, it is important to recognize that the provision of this care can be – and is being – made without them. From the standpoint of social policy, the prudent course would appear to be to do what is necessary to ensure that the aged receive proper care, irrespective of who provides that care, rather than becoming bogged down in fretting over whether there will be enough close kin around to perform the functions traditionally thought to attach to the proper role of kin in such matters. It makes more sense to put the emphasis on how to meet these needs rather than on the character-istics of those assisting in the achievement of such a goal. All of this leads to consideration of the need to provide an appropriate physical and social setting for such provision, which is something that will be addressed in Chapter 7.

REFERENCES

Aaron, H. J., B. P. Bosworth & G. T. Burtless 1989. *Can America Afford to Grow Old? – Paying for Social Security*. Washington, DC: Brookings Institution.

Aaron, H. J. & W. B. Schwartz 1984. *The painful prescription: rationing hospital care*. Washington, DC: Brookings Institute.

Australia, Bureau of Transport and Communications Economics 1989. Cost of road crashes in Australia, 1988, *Information Sheet 1*. Canberra: B.T.C.E.

Australia, Social Welfare Policy Secretariat 1984. *The impact of population changes on social expenditure: projections from 1980–81 to 2021.*

Australian Bureau of Statistics 1988. *Causes of death: Australia, 1986.* Canberra: Commonwealth Government Printer.

Australian Institute of Health 1988. *Australia's health*. Canberra: Australian Government Publishing Service.

Barer, M., I. R. Pulcins, R. G. Evans, C. Hertzman, J. Lomas & G. M. Anderson 1988. Diagnosing senescence: trends in medical service utilization by B.C.'s [British Columbia's] elderly. Mimeo.

Bartlema, J. D. 1987. *Developments in kinship support networks for the aged in the Netherlands*. Tilburg: Reeks Sociale Zekerheidswetenschap, Katholieke Universisteit Brabant.

Bates, E. 1980. Talk presented on Australian Broadcasting Commission radio program, The Body Program, February 10.

Benet, S. 1974. *The long-living people of the Caucasus*. New York: Holt, Rinehart & Winston.

Berkman, L. F. 1985. The relationship of social networks and social support to morbidity and mortality. In *Social support and health*, S. Cohen and S. L. Syme (eds.). New York: Academic Press.

Bloom, B. S. & P. D. Kissick 1980. Home and hospital cost of terminal illness. *Medical Care* 18(5), 560–4.

Brody, J. A. 1985. Prospects for an ageing population. *Nature*, June 6, 463–6.

Brody, J. A. 1987. The best of times/the worst of times: ageing and dependency in the 21st century. In *Ethical dimensions of geriatric care*, Stuart F. Spicker, Stanley R. Ingman & Ian R. Lawson (eds.). Dordrecht, Netherlands: D. Reidel.

Brody, J. A., D. B. Brock & T. F. Williams 1987. Trends in the health of the elderly population. *Annual Review of Public Health* 8, 211–34.

Broe, G. A. 1986. Head injury – the nature and extent of the problem. Headway Conference, Sydney.

Callahan, D. 1987. *Setting limits – medical goals in an aging society*. New York: Simon & Schuster.

Carter, J. 1981. *States of confusion: Australian policies and the elderly confused*. Social Welfare Research Centre Reports and Proceedings, no. 4. Kensington, N.S.W.: University of New South Wales.

Cass, B., F. Gibson & F. Tito 1988. *Social security review. Towards enabling policies: income support for people with disabilities*. Canberra: Australian Government Publishing Service.

Castles, F. G. 1982. The impact of parties on public expenditure. In *The impact of parties: politics and policies in democratic capitalist states*, F. G. Castles (ed.). Beverly Hills, Calif.: Sage Publications.

Chassin, M. R. & R. H. Brook 1988. Research reported in 'Doctor, is this operation necessary?'. *Rand Research Review* **12(3)**, 1–3.

Cherlin, A. J. & F. F. Furstenberg, Jr 1986. *The new American grandparent – a place in the family, a life apart*. New York: Basic Books.

Crown, W. 1984. The prospective burden of an aging population. In *Of current interest from the Policy Center on Aging* **4(1)**. Waltham, Mass.: Florence Heller Graduate School, Brandeis University.

Daly, H. E. 1986. Population growth and economic development: policy questions. *Population and Development Review* **12(3)**, 582–5.

Davis, K. 1988. Our idle retirees drag down the economy. *New York Times*, October 18.

Day, A. T. 1985a. *'We can manage' – expectations about care and varieties of family support among persons 75 years of age and over*. Melbourne: Institute of Family Studies.

Day, A. T. 1985b. *Who cares? Demographic trends challenge family care of the elderly*. Washington, DC: Population Reference Bureau.

Day, A. T. 1987. Characteristics of the aged in institutions – current patterns and future prospects. Paper presented at International Union for the Scientific Study of Population seminar, Vaucresson, France, October 6–9.

Day, A. T. 1991. *Remarkable survivors: insights about successful aging among women*. Washington, DC: Urban Institute Press.

Day, L. H. 1978. Government pensions for the aged in 19 industrialized countries: demonstration of a method for cross-national evaluation. In *Comparative studies in sociology*, Richard Tomasson (ed.), Vol. 1, 217–33.

Day, L. H. 1984. Death from non-war violence: an international comparison. *Social Science and Medicine* **19(9)**, 917–27.

Doty, P. 1986. Long-term care for the elderly provided within the framework of health care schemes. Report of the Permanent Committee on Medical Care and Sickness Insurance, Health Care Financing Administration (U.S.A.), presented at 22nd General Assembly, International Social Security Association, Montreal, September 2–12.

Duggan, J. M. 1989. Technological triumph – trivial pursuit. *Australian and New Zealand Journal of Medicine* **19**, 506–8.

E.P.A.C. (Australia: Economic Policy Advisory Council) 1986. *Growth in Australian social expenditures*. Council Paper No. 17.

E.P.A.C. 1988. *The economic effects of an ageing population.* E.P.A.C. Paper No. 29. Canberra: Australian Government Printing Service.

Fisher, C. R. 1980. Differences by age groups in health care spending. *Health Care Financing Review* 1(4), 65–90.

Fries, J. F. 1980. Aging, natural death, and the compression of morbidity. *New England Journal of Medicine* 303, 130–5.

Fuchs, V. R. 1984. Sounding board: the 'rationing' of medical care. *New England Journal of Medicine* 311, 1572–3.

Garber, A. M. 1987. Long term care, wealth, and health of the disabled elderly living in the community. Working Paper No. 2328, National Bureau of Economic Research (abstracted in *N.B.E.R. Reporter*, Fall).

Gibson, D. E. 1989. Advancing the dependency ratio concept and avoiding the Malthusian trap. *Research on Aging* 11(2), 147–57.

Gonnot, J.-P. 1990. Demographic, social, and economic aspects of the pension problem: evidence from 12 countries. *Popnet*, No. 17, 3–10. Laxenburg, Austria: International Institute for Applied Systems Analysis.

Guralnick, J. M. & E. L. Schneider 1987. Prospects and implications of extending life expectancy. In *Technological prospects and population trends*, Thomas J. Espenshade & George J. Stolnitz (eds.). A.A.A.S. Selected Symposium, No. 103. Boulder, Colo: Westview Press.

Guralnick, J. M., D. B. Brock & J. A. Brody 1987. The changing demography of the elderly in the United States. In *Advanced geriatric medicine*, F. I. Caird & J. Grimley Evans (eds.). Bristol, Conn.: Wright.

Higgins, L. C. 1989. Care-rationing hits at local level. *Medical World News* 24, 21–3.

Hook, S. 1987. In defense of voluntary euthanasia. *New York Times*, March 1.

I.N.E.D. (Institut National d'Etudes Demographiques) 1988. Dix-septieme rapport sur la situation demographique de la France. *Population* 43(4–5), 727–98.

Katz, S., L. G. Branch, M. H. Branson *et al.* 1983. Active life expectancy. *New England Journal of Medicine* 309, 1218–24.

Kay, D. W. K. 1980. Epidemiology of mental disorders amongst the aged in the community. In *Handbook of mental health and ageing*, J. E. Birren & R. B. Sloane (eds.). Englewood Cliffs, N.J.: Prentice-Hall.

Keyfitz, N. 1989. Ageing is not the whole pension problem, *Popnet*, No. 16, 5–8. Laxenburg, Austria: International Institute for Applied Systems Analysis.

Keyfitz, N. & W. Flieger 1968. *World population.* Chicago: University of Chicago Press.

Kuhse, H. 1987. Voluntary euthanasia in the Netherlands. *Medical Journal of Australia* 147, 394–6.

Kuhse, H. & P. Singer 1988. Doctors' practices and attitudes regarding voluntary euthanasia. *Medical Journal of Australia* 148, 623–7.

LaLonde, M. 1974. *A new perspective on the health of Canadians.* Ottawa: Department of National Health and Welfare.

Lamm, R. D. 1984. Long time dying. *The New Republic*, August 27, 20–3.

Levinsky, N. G. 1984. Sounding board: the doctor's master. *New England Journal of Medicine* 311, 1573–5.

Litwak, E. 1985. *Helping the elderly: complementary roles of informal networks and formal systems.* New York: Guilford Press.

Litwak, E. & P. Messeri 1989. Organizational theory, social supports, and mortality rates: a theoretical convergence. *American Sociological Review* 54(1), 49–66.

Lubitz, J. & R. Prihoda 1983. Use and costs of Medicare services in the last years of life. *Health, United States, 1983.* Washington DC: U.S. Department of Health and Human Services.

McCall, N. 1984. Utilization and costs of Medicare services by beneficiaries in their last year of life. *Medical Care* 22(4), 329–42.

McGregor, M. 1989. Technology and the allocation of resources. *New England Journal of Medicine* 320, 118–20.

Maddox, G. L. 1987. Aging differently. *The Gerontologist* 27(5), 557–64.

Manton, K. G. 1987. The population implications of breakthroughs in biomedical technologies for controlling mortality and fertility. In *Technological prospects and population trends*, T. J. Espenshade & G. J. Stolnitz (eds.), A.A.A.S. Selected Symposium, No. 103. Boulder, Colo: Westview press.

Manton, K. G. & K. Liu 1984. The future growth of the long term care population: projections based on the 1977 Nursing Home Survey and the 1982 Long Term Care Survey. Prepared for presentation at Third National Leadership Conference on Long-Term Care Issues, Washington, DC, March.

Manton, K. G. & B. Soldo 1985. Dynamics of health changes in the oldest old: new perspectives and evidence. *Milbank Memorial Fund Quarterly/ Health and Society* 63(2), 206–85.

Marshall, V. W. (ed.) 1980. *Aging in Canada.* Don Mills, Ont.: Fitzhenry & Whiteside.

Mathers, C. & R. Harvey 1988. *Hospital utilisation and costs study*, Vol. 2, *Survey of public hospitals and related data.* Australian Institute of Health, Canberra: Australian Government Publishing Service.

Nathanson, C. A. & A. D. Lopez 1987. The future of sex mortality differentials in industrialized countries: a structural hypothesis. *Population Research and Policy Review* 6, 123–36.

N.C.H.S. (U.S. National Center for Health Statistics) 1987. Use of nursing homes by the elderly: preliminary data from the 1985 National Nursing Home Survey. *Advance Data from Vital and Health Statistics*, no. 135. D.H.H.S. publication no. (P.H.S.) 87–1250. Public Health Service, Hyattsville, Md, May 14.

Ogawa, N. 1985. Consequences of mortality change on aging. N.U.P.R.I. Reprint Series No. 20. Tokyo: Nihon University Population Research Institute.

O.E.C.D. (Organisation for Economic Co-operation and Development) 1987. *Financing and delivering health care.* Paris: O.E.C.D.

O.E.C.D. 1988a. *Ageing populations: the policy implications.* Paris: O.E.C.D.

O.E.C.D. 1988b. *The future of social protection.* O.E.C.D. Social Policy Studies, no. 6, Paris: O.E.C.D.

Pamuk, E. R. 1985. Social class inequality in mortality from 1921 to 1972 in England and Wales. *Population Studies* 39(1), 17–31.

Pamuk, E. R. 1988. Social class inequality in infant mortality in England and Wales from 1921 to 1980. *European Journal of Population* **4**(1), 1–21.

Porterba, J. M. & L. H. Summers 1987. Public policy implications of declining old-age mortality. In *Work, health, and income among the elderly*, Gary T. Burtless (ed.). Washington DC: Brookings Institution.

Pullum, T. W. 1982. The eventual frequencies of kin in a stable population. *Demography* **19**(4), 549–65.

Rabin, D. L. & P. Stockton 1987. *Long-term care for the elderly: a factbook*. New York: Oxford Unversity Press.

Riley, J. C. 1990. Morbidity trends in four countries: the risk of being sick. Paper presented at annual meeting, Population Association of America, Toronto, May 2–5.

Riley, M. W. & K. Bond 1983. Beyond ageism: postponing the onset of disability. In *Aging in society: selected reviews of recent research*, M. W. Riley, B. B. Hess & K. Bond (eds.). Hillsdale, N.J.: Lawrence Erlbaum Associates.

Ringen, S. 1987. *The possibility of politics: a study in the political economy of the welfare state*. Oxford: Oxford University Press.

Roos, N. P., P. Montgomery & L. L. Roos 1987. Health care utilization in the years prior to death. *Milbank Quarterly* **65**(2), 231–54.

Rosenmayr, L. 1981. Help and self-help: some problems of aging policy. In *The situation of the elderly in Austria*, Charlotte Nusberg (ed.). Washington, DC: International Federation on Aging.

Rosenmayr, L. & E. Kockeis 1968. Propositions for a sociological theory of aging and the family. *International Social Science Journal* **15**, 410–26.

Rosenwaike, I. 1985. A demographic portrait of the oldest old. *Milbank Memorial Fund Quarterly/Health and Society* **63**(2), 187–205.

Rowland, D. T. 1984. Old age and the demographic transition. *Population Studies* **38**(1), 73–87.

Rowland, D. T. 1991. *Ageing in Australia*. Melbourne: Longman Cheshire.

Sagan, L. A. 1987. *The health of nations*. New York: Basic Books.

Sawitz, E. *et al.* 1988. The use of in-hospital physician services for acute myocardial infarction: changes in volume and complexity over time. *Journal of the American Medical Association* **259**(16), 2419–22.

Schatzkin, A. 1980. How long can we live? A more optimistic view of potential gains in life expectancy. *American Journal of Public Health* **70**(11), 1199–1200.

Scitovsky, A. A. 1984. 'The high cost of dying': what do the data show? *Milbank Memorial Fund Quarterly/Health and Society* **62**(4), 591–608.

Scotto, J. & L. Chiazze 1976. *Third national cancer survey: hospitalization and payments to hospitals*. Part A: Summary. U.S. Dept. of Health, Education, and Welfare publication no. NIH–76–1094. Bethesda, Md: National Institutes of Health.

Shulkin, D. J. 1988. Letter. *New England Journal of Medicine* **319**(19), 1291.

Stamler, J. 1985. Coronary heart disease: doing the 'right things'. *New England Journal of Medicine* **312**(16), 1053–5.

Statistics Sweden 1982. Unpublished data from 1980–81 Survey of living conditions. Available from U.N. Statistical Office 1989.

Stone, A. A. 1985. Sounding board: law's influence on medicine and medical ethics. *New England Journal of Medicine* 312, 309–12.

Suzman, R. & M. W. Riley 1985. Introducing the 'oldest old'. *Milbank Memorial Fund Quarterly/Health and Society* 63(2), 177–86.

Thurow, L. C. 1984. Sounding board: learning to say 'no'. *New England Journal of Medicine* 311, 1569–72.

Tilquin, C. *et al.* 1980. The physical, emotional, and social condition of an aged population in Quebec. In *Aging in Canada*, Victor W. Marshall (ed.). Don Mills, Ont.: Fitzhenry & Whiteside.

Timmer, E. J. & M. G. Kovar 1971. Expenses for hospital and institutional care during the last year of life for adults who died in 1964 or 1965. *Vital and Health Statistics*, Series 22, no. 11. Hyattsville, Md: U.S. Dept. of Health, Education, and Welfare.

Torrey, B. B. 1988. Assets of the aged: clues and issues. *Population and Development Review* 14(3), 489–97.

Troyer, H. 1988. Review of cancer among four religious sects: evidence that life-styles are distinctive sets of risk factors. *Social Science and Medicine* 10, 1007–17.

U.N. (United Nations) 1981, 1985. *Demographic Yearbook*. New York: United Nations.

U.N. Statistical Office 1989. Disability Statistics Data Base 1975–86 (unpublished data).

U.S. Department of Health, Education, and Welfare, Office of Research and Statistics 1967, 1975. *Social security programs throughout the world*. Washington, DC: Government Printing Office.

U.S. (United States of America), Health Interview Survey.

Verbrugge, L. 1988. Arthritis and disability in older adults. Paper presented at annual meeting of Population Association of America, New Orleans.

Vossen, A. & T. Janssen 1987. Countering socio-economic consequences of ageing: the case of the Netherlands. Paper presented at European Population Conference, Jyvaskyla, Finland, June.

Warren, M. D. 1987. The prevalence of disability: measuring and estimating the number and the needs of disabled people in the community. *Public Health* 101, 333–41.

Wilensky, H. L. 1975. *The welfare state and equality*. Berkeley/Los Angeles: University of California Press.

Wolf, D. A. 1986. Kinship and family support in aging societies. Working Paper 86–81. Laxenburg, Austria: International Institute for Applied Systems Analysis.

Wolf, D. A. 1988. Family structure and caregiving portfolios. Working paper. Laxenburg, Austria: International Institute for Applied Systems Analysis.

Wrigley, J. M. & C. B. Nam 1987. Underlying versus multiple causes of death: effects on interpreting cancer mortality differentials by age, sex, and race. *Population Research and Policy Review* 6(2), 149–60.

Wu, C. 1986. Contribution of the elderly population to society: China's perspectives and experiences. Paper presented at United Nations International Symposium on Population Structure and Development, Tokyo, September 10–12.

Yeo, S. 1985. *Recent advances in the treatment of patients with spinal cord injury*. Sydney: Acute Spinal Unit, Royal North Shore Hospital.
Zapf, W. 1984. Welfare production: public versus private. *Social Indicators Research* 11(3), 263–74.

4

THE CHALLENGE OF NUMERICAL DECLINE AND OLDER AGE STRUCTURE:
Part 2 Households, labor force, economic conditions and behavior

INTRODUCTION

The age-relatedness of certain physical conditions and stages of life, together with the modern welfare state's provision of certain types of social services (such as age pensions and supporting mothers' benefits) on the basis of age or age-related condition, permits a high degree of predictability about some of the consequences of the demographic conditions to be expected in the countries under consideration – those relating to finances, health and the need for care, for example. But about others, particularly those relating to behavior, there can be much less certainty. The experience to go on is simply too limited. Age structures as old, and growth rates as low, as those expected here have been exceedingly rare and short-lived; while zero (or negative) population growth rates originating specifically in long-term control over fertility (rather than in emigration or increased mortality) are unique to the present age.

Nonetheless, there are some certainties, if only at a high level of generality: (a) the range of possibilities is broad and (b) what happens will depend only slightly on demographic condition. Between a population's characteristics and its conditions of life there is no one-to-one causal relation. Demographic size and composition will set limits, but these will not, except at the very extremes of over-population and under-population, be the ultimate determinants of the conditions of human life. In comparison with their alternatives, the conditions expected for the populations under consideration here could make the 'good life' more attainable, but they will not, in themselves, produce it. Life under these conditions could be meagre or bountiful, violent or peaceful, miserable or

84

happy. But whatever the demographic circumstances – again, short of the extremes – conditions of life will be more a consequence of institutional structures and social attitudes and policies than of any population characteristics as such.

So far as conditions of life are concerned, the significance of the demographic circumstances envisaged may well lie more in the sheer sizes of these populations once they have ceased to grow than in their respective age/sex structures. On the assumption of fertility at exact replacement level after the year 2020, three out of four of these countries will experience further numerical increases. These will range from less than 5 per cent (in six countries) to more than 20 per cent (in eight countries), with a median increase of 14 per cent. Such growth – not to mention the unquestionably more significant increase of some 40 per cent expected in world population as a whole over the next couple of decades (U.N. 1989, Table 2.1) – when combined with the several exceedingly detrimental (often irreversible) processes already well under way in the natural environment (see, e.g., McKibben 1989, Smil 1990), can be expected to bring changes in human conditions and lifestyles scarcely imaginable in terms of present perspectives. Keeping these broader, all-pervasive demographic, and especially environmental, changes ever in mind, what will be addressed here are those social arrangements and items of behavior – namely, family and household composition, the labor force, certain economic conditions, and behavior and politics – which are generally considered particularly susceptible to modification in consequence of the sorts of demographic developments envisaged.

FAMILY AND HOUSEHOLD COMPOSITION

Because most people in these low-mortality populations die in old age, the transition to older age structures will result in a higher proportion widowed. The extent of this will depend on the level of male/female mortality differences and the direction and magnitude of age differences between husbands and wives. Any lessening of these differences, or a change in marriage patterns so that the wife, rather than the husband, was more often the older partner, would reduce both the extent and the duration of widowhood. It would not completely compensate for the effect of age, however; there is no way of getting around the fact that older age structures will inevitably be characterized by more widowhood.

Older age structures can also be expected to have higher proportions who have experienced marital dissolution through divorce or separation, too, if only because of longer exposure to the risks of such an occurrence. Changes in the availability (legal, financial, or normative) of divorce could for a time result in higher proportions of divorcees at the younger ages, but the overall effect of aging as such would be to increase the proportion of the population who would have experienced marital dissolution in some form – most of it the result of death, but some the result, instead, of divorce and separation.

Marital dissolution has significant implications for household composition. Especially at the older ages, when children have grown up and left home, is it likely to be associated with living either in a single-person household or an institution. This is of no little moment to human wellbeing, even if the association can take more than one direction. The composition of his or her household is a major determinant of the amount and type of social support an older person is likely to have available. Whether the older person lives alone, with a spouse, or with an adult child both reflects and conditions a variety of economic and social factors with important bearing on the nature of caregiving relationships. Living with a spouse, for example, generally maximizes the opportunity for receipt of substantial long-term personal care, while living alone, in contrast, requires a minimum capacity for self-help. Living alone also means that any assistance has to come from outside the household and that there is, therefore, a risk of intermittent supervision, of delays in the receipt of needed help, of inconvenience to the caregiver(s). While living with an adult child, instead, can permit close supervision and immediate reaction to need, it can also involve intergenerational relations at close quarters and potential stress arising out of the older person's dependency, role reversal between parent and child, and, for the primary caregiver, the possibility of competing loyalties and demands between husband and children on the one hand and older parent on the other. Where conditions permit, most older people seem to prefer to live on their own, maintaining with their families a relation of 'intimacy, but at a distance.' Where this is the preferred arrangement, living with an adult child is apt to reflect not so much a reasoned choice as a condition of physical or financial dependency on the part of the aged parent. And because it is so often tied in with limited economic resources, it is also apt to reflect considerable financial strain and residential crowding.

In her study of a nationwide random sample of ever-married white American women 77 to 87 years of age, A. T. Day (1991) found a strong desire to be independent and not a burden coupled with uncertainty and no little anxiety about what will happen if stamina declines and maintaining a separate household is no longer feasible. The alternatives – living with children or moving to an institution – appeared to be regarded as almost equally unsatisfactory. Nevertheless, given the association between advancing age and declining capacity for independent activity, the transition to older age structures is likely to increase the need for various types of shared households so that older people can, if it becomes necessary, change from living largely without assistance in their own households to living with possibly a substantial amount of assistance in a more protective environment.

Not even at the extreme, however, would this be likely to involve more than a small minority of the aged; the exact proportion depending, in all likelihood, far less on age itself than on the economic and social setting. Fewer than 9 per cent of the 77 to 87-year-old women in A. T. Day's study were living in institutions. Of the rest, 55 per cent were living alone, 21 per cent with only their husbands, 16 per cent with one or another of their children, and 3 per cent with a brother or sister. The importance of kinship in living arrangements is seen in the fact that fewer than 2 per cent were living only with non-kin. The variety of household types was considerable: in all, 33 different combinations, which, if nothing else, suggests the existence of a substantial reservoir of adaptive capacity among these older women. One of the respondents was living with her husband, grandson, and grandson's friend; another with her sister, son-in-law, and grandchild; five were living only with grandsons, and 10 only with friends.

THE LABOR FORCE

Under the demographic conditions envisaged, those of working age will be a somewhat higher proportion of the total, and they will have a somewhat older age structure. But by how much is in either instance a matter, partly, of arbitrary definition. For one thing, there is the arbitrariness about what constitutes 'working.' Some of the most important production of goods and services occurs outside the marketplace. Even in highly industrialized societies with well-developed social welfare systems, the family, in addition to being important as an emotional community, is also important as an arena of production (Ringen 1987, 123). In terms of actual function, for example, a German study (Zapf 1984) finds that private households are that country's biggest producers of transportation, food, and laundry services; that about a third of that country's households are engaged in agriculture (working their own fields or kitchen gardens), that about a third are also engaged in home-building, as well as large numbers in various kinds of repair services; and that two-thirds are engaged in providing neighborhood help. The number of old, sick, and handicapped persons taken care of in households far exceeds the number who could ever be admitted to institutions.

Nevertheless, in conformity with the usual practice, 'working' will for present purposes be taken to mean participation in the labor market: either working for pay or profit or looking for such work. Taking the United Kingdom as an example, if we assume working age to range from 18 to 65, the expected changes in age structure would, at the one extreme of the World Bank's assumption of fertility at exact replacement level after the year 2020, reduce the proportion of working age by 3.1 percentage points from what it was in 1985. Under the other extreme assumption, that of continuation of 1985 fertility levels, it would reduce this proportion by 3.5 percentage points. If, on the other hand, we take 20–69 as our 'working' age, in recognition of the fact that a longer period of schooling and training is now characteristic of these economies (irrespective of whether it is actually necessary to their functioning), and that changes in health and organization are likely to permit participation in the labor market at more advanced ages, the decrease would be some 1.6 percentage points under the replacement-level fertility assumption, and only 0.7 percentage points under the assumption of a continuation of 1985 fertility. The situation in Sweden would be much the same: decreases of 3.0 and 3.7, respectively, with working age taken as 18–64, and of 2.4 and 1.4, respectively, with working age taken as 20–69. At either of the extreme assumptions about fertility, the proportion of working age would decline, but it would not do so by much.

However, this delineation of 'working' age is an exceedingly crude indicator of the number who will actually be either in the labor market or available for entry into it. There is much more than age structure to the creation of a labor force. Besides varying patterns of schooling and retirement, there are: the willingness of women in general, and the mothers of small children in particular, to seek employment outside the home; the patterns of marriage, divorce, and separation – which can greatly affect the numbers of women who will be self-supporting; and then, of course, economic conditions themselves, like inflation, employment opportunities, and patterns of consumption. Demographic circumstances may set the limits, but the range of variation within these limits can be very broad.

Some, looking only at the crudely demographic side of the equation, and reasoning on the basis merely of the ratio of numbers of elderly to numbers of working age, foresee low fertility leading eventually to labor shortages in several of these countries. Quite apart from its incorporating the error, already dealt with in Chapter 4, of assuming that all elderly are 'dependent' upon those in the working ages, and that these demographic changes will bring no compensating reductions in the 'dependency' of youth, such a view overlooks the possibilities for increasing labor market participation through, for instance, the introduction of greater flexibility in scheduling, more part-time work, and improved working conditions. Even where it is relatively high, the rate of female participation in the labor force is well below that of men in the same age categories: at age 40–44, for example, it is 22 per cent lower in the U.S.A., and 31 and 32 per cent, respectively, in the United Kingdom and France (I.L.O. 1986, Table 1). Only in the Scandinavian countries and Finland is it no more than 10 per cent lower, and here the narrower gap is at least partially explained by differences in definition. It will doubtless be possible to fill any work force gap resulting from low fertility by drawing further on the pool of 'marginal' workers: women, especially, but also the disabled, youth, and the retired. In fact, continued low fertility, by limiting the proportion of women with young children and thereby freeing more of them for participation in the labor market (assuming they chose to do this with their time) could actually increase the size of the labor force.

Much the same can probably be said, as well, of labor mobility. Here, the fact that the work force will be somewhat older is often deemed a matter of some economic significance. Compared with the situation in 1985, stabilization of the age structures in these populations through continuation of 1985 fertility levels would generally increase the proportions of 50+ years of age among those of 'working' age (assuming this to be 20–64) by less than 10 percentage points (although in the Netherlands, where a period of comparatively high fertility in the pre-and post-World War II period was followed by one of particularly rapid decline, the increase would be as high as 13 percentage points). Were these populations to achieve stability with fertility at replacement levels, instead, the comparable increases would in most instances be less than 5 percentage points (estimated from 1984 Swedish life table in U.N. 1985, Table 36). Young people, having received the most up-to-date education and training, are viewed by some as being both better able and more willing to change their occupations and their places of work (Wander 1977). But whether or not this is, indeed, the case, the differences between what exists today in work force age structures and what can be expected with the continuation of low fertility is not very great. Moreover (and more important), the causal factor in such mobility is surely social position, not age. Given other circumstances, greater mobility might be found at other ages (Pryor 1977). And what, too, of mobility itself? Is it invariably desirable – as economic neo-orthodoxy would have us believe? Might it not give rise to social and psychic costs – arising, for example, out of the loss of networks of kin and friends – that more than outweighed any economic benefit it might occasion?

The generally older work force in these countries would presumably be more experienced and more skilled, as well as more in possession of economically useful attitudes. But if some skills, and particularly some attitudes (such as discipline and a sense of time) are of universal value in an industrial economy, it is possible that others, similarly associated with age and experience, are not. To the extent that economic development depends on changing the distribution of jobs by occupational category, it is conceivable that it could be retarded more than advanced by the possession of certain skills and attitudes found in greater profusion among older workers; for it is the older worker who would have the most to lose – economically, socially, and emotionally – by such changes. Yet, the economically undesirable effects of these workers' inhibitions could probably be largely offset by the greater employment of such currently 'marginal' workers as students, the disabled, mothers of young children, or older women with little prior work experience – none of whom would presumably have the same degree of interest that more experienced workers would have in maintaining a particular occupational or work process status quo. If constraints on the supply or behavior of labor under the demographic conditions to be expected in these countries do eventuate, they will be not demographic in nature but essentially institutional and social.

ECONOMIC CONDITIONS

While a different pattern of economic conditions in some of these countries would undoubtedly be preferable from the standpoint of the conservation of resources or the enhancement of the quality of human life (see, e.g., Mishan 1967, L. H. Day 1971), there is nothing in the expected demographic conditions themselves that would seem inevitably productive of any particular economic change or condition. Economic change, stability, or decline; full employment or high levels of unemployment or underemployment (Kelley 1972 and 1989, Spengler 1972, Robinson 1972); economic activities that stress the private sector and those that stress the public; those that are destructive of environmental amenity and those that are not – all are possible. To the extent that it exists at all, demographic determinism in economic matters is notably lacking in specificity.

It is doubtful whether industrialism, given its dependence on a high rate of consumption of both renewable and nonrenewable resources, can continue much longer in anything approaching its present form; or, given its deleterious consequences for the environment, whether it ought to be allowed to, even if it could. Biologists, ecologists, occasionally geologists and chemists, have been raising these questions for a number of years. More recently, a few economists have joined in (see, e.g., Boulding 1966, Daly 1971 and 1977, Schumacher 1973), and now, even more recently, so, also, have some among the political elites in industrialized countries (see, e.g., World Commission on Environment and Development 1987). The idea that an older age structure might better lend itself to undertaking the kinds of economic changes these conditions call for will be developed in Chapters 6 and 7. For purposes of the present discussion, however, the continuation of industrialism will be assumed.

Investment

If business confidence depends on population growth, then stability or declines in numbers will, by definition, lessen that confidence and the investment it would presumably have encouraged. Whether businessmen, faced with the absence of numerical increases could persist in making their confidence dependent upon such increases is assuredly open to question. But it does not appear that they would need to: business confidence has been low during periods of rapid population increase and, judging from Scandinavian experience, high during periods of low population increase. The cessation of population growth does not, apparently, preclude an expanding economy. Whether an expanding economy is a desirable social goal is, of course, another matter. At the least, one ought to ask: Expansion of what aspects of the economy? Where? By what means? How fast? And at what cost (and to whom)?

Compared with banks, insurance companies, and businesses themselves, individuals are of little importance as sources of capital in industrial economies. It is possible, however, that they might play a relatively greater role as a source of genuine 'risk' capital. But where this is the case, so far as purely demographic conditions are concerned, the rate of capital formation will more likely be a response to purely local conditions, such as internal migration and unemployment rates or even family size among the owners of capital, than to the generality of demographic conditions themselves.

Inflation

It has been suggested (Spengler 1972) that the cessation of population growth will promote inflation because: (a) it will bring a relative shortage of labor, (b) its labor force, being older, will be less mobile, (c) much of the increased demand it makes for labor will be in the service industries, where productivity has traditionally been low, and (d) the increased political power of older people stemming from their increased proportion of the population will fuel the inflationary spiral by forcing governments to supplement pensions and other payments to them as compensation for the losses they will have incurred through inflation. None of these eventualities is a certainty. Nor have these hypotheses received much support in what has occurred during the nearly two decades since they were made. But even if they were certainties, other attributes of the expected demographic conditions would seem to be conducive to less, not more, inflation – and to be so to an extent sufficient to counterbalance these presumed inflationary pressures. For one thing, if a pattern of slight annual fluctuations in birthrates were to continue, the resulting smoother age distribution would lead to less fluctuation, at least at the national – and possibly, regional – level, in demand for schooling, jobs, and housing (the kind of fluctuation that originates in sudden swings in the range of formation of new families and households). The greater certainty in forecasting permitted by such a population could also result in less speculation, certainly in real estate but also in other business activity, as well, for there would be no specifically demographic encouragement in such a population structure to overcapitalization or to unwise investment on the assumption that a larger potential market, flowing inevitably from an increase in population, would somehow compensate for possible errors in judgment.

Yet, overall, the role of demographic factors in promoting or retarding inflation is surely minimal. Primary importance must lie, firstly, in government decisions concerning such things as taxation, trade, public expenditure, wage and price controls, and forced savings and transfer payments for retirement, health, welfare, and recreation; and, secondly, in the decisions of the larger businesses (and labor unions) concerning remuneration, postponed wages payment plans, investment, and the introduction of labor-saving devices and techniques. Only in the case of sudden population increases from immigration, and then very largely only in the case of housing markets (e.g., Joske 1989, Nielsen 1989), do demographic factors appear to have much importance in the promotion of inflation.

Employment

Probably much the same also holds for the relation between these demographic conditions and employment levels. If slower or negative growth rates and older age structures did, indeed, produce labor shortages (which, as already noted, seems most unlikely), this would conceivably reduce unemployment, at least in the short run. But, as with the question of inflation, there is surely at work a host of other factors of far greater significance.

However, we could expect these older age structures to foster some increases in employment in certain of the caring services, nonprofessional as well as professional; and also in some employments – those associated with public transportation or health counseling, for example, that, while focused less exclusively on the needs of the elderly, are of particular importance to them.

As it happens, many of the services and facilities conducive to the wellbeing of the aged are also conducive to the wellbeing of the rest of the society, perhaps particularly to that of children and young adults. Some that come immediately to mind are: public transportation, safe places to walk, parks and plazas in which to meet and be with people on an informal basis, low-cost rental housing close to shops and other public facilities, small-scale shops that sell products in small quantities and permit personal contact between shopper and shopkeeper, walk-in health and counseling services, even clean public washrooms. The tendency to pit the cost of providing for the aged against that of providing for other members of society – implying in the process that there is a pervasive divergence of interests between the aged and the rest of the society, and that serving those of the one must necessarily be to the neglect of those of the other (see, e.g., Preston 1984; Thomson, 1989) – is generally misplaced. Not only are the actual amounts involved often quite modest in comparison with those for other items of government expenditure (a couple of prime examples being: military preparedness and space exploration, at the national level, and highway construction and maintenance, at the local), but, in most instances, the benefits extend well beyond the confines of the aged themselves. Divergencies of interest do exist, but on most of these issues, their importance would seem to be more than outweighed by the congruencies.

BEHAVIOR AND POLITICS

On the assumption that older people tend to be more conservative, it has been widely assumed that societies with older age structures would also be more conservative. Some years ago, for example, a well-known economist–demographer averred that a society with the older age structure of a stationary population 'would not be likely to be receptive to change and indeed would have a strong tendency towards nostalgia and conservatism;' and a famous French demographer has characterized such a population as one of 'old people ruminating over old ideas in old houses.'

At first glance, the assumption does not seem unreasonable. Apart from occupying different statuses and playing different roles in the present, older people will have behind them a lifetime of experiences in many ways different from those of the rest of society; experiences out of which could be expected to emerge a measure of difference in basic values, attitudes, and world views between older people and the rest of society. In concert with the tendency of the healthy personality to slough off unpleasant memories, such differences in outlook could be expected to lead older people to take a rather more positive view of the past, and, by way of contrast, a rather les positive view of the present. (Which, of course, begs the question of whether – to those with experience of both – the past might not, indeed, have been truly better than the present.) That the past would also have been for the older person a time of possibly better health, more stamina, unimpaired sexual drive and the like would merely heighten its relative appeal.

Then there is the matter of 'attitude flexibility,' which is widely viewed as steadily decreasing with age because of (summarized in Krosnick & Alwin 1989, 417): (a) a declining capacity for information processing and memory as a result of declines in energy and the loss of brain tissue, (b) the accumulation, with experience, of 'attitude-relevant' knowledge that serves as a source of psychological stability, (c) less consideration of attitude-challenging information because of social disengagement and decreased interest in events distant from one's immediate life, and (d) increased support for one's attitudes through the accumulation of friends with similar attitudes.

There are some difficulties with this line of reasoning, however. For one thing, the form of the argument itself comes dangerously close to committing the logical fallacy of 'composition:' assuming that what is true of a part is also true of the whole (Hamblin 1970). More important is the paucity of empirical support, and the suggestion that whatever association there might be could be in the opposite direction. In terms of demonstrated willingness to introduce changes in economic behavior, education, and the status of women, for example, the world's most conservative societies have tended to be those with the youngest – not the oldest – age structures. Moreover, before the 1973 abortion law reform in the U.S.A., public opinion polls showed older women *more* – not less – willing than younger women to remove legal restrictions on access to abortion (Blake 1971). Was this lesser receptivity to change? In 1968, the *New Republic* magazine continued its long-standing practice of rating the 'progressivism' of members of the American Congress. On the basis of the twelve 'key' Senate votes selected, 'progressivism,' as adjudged by this, at the time, self-styled 'liberal' magazine, was indeed associated with (relative) 'youth:' the median age of those with no more than three 'unfavorable' votes being four years younger than that of those with no more than three 'favorable' votes. But on the one vote that reflected attitudes toward the

Vietnam War, the 'hawks' and the 'doves' had the same median age, and there was no difference between them in distribution by age. Twenty years earlier, in the 80th Congress, a Congress in which 'liberals' and 'conservatives' were probably more clearly distinguished from one another than at any time since, the Senate ratings again showed a difference of four years between the median ages of the 'liberals' and the 'conservatives'. Only, this time, the order was reversed: it was the 'liberals' who were older (L. H. Day 1978, 28–9).

European opinion polls on issues of some arguable importance offer data no less uncertain as to the significance of age. Stated unwillingness to accept the risks of nuclear energy, for instance, while negatively associated with age in France and Germany, bore no consistent association with age in Belgium, the Netherlands, Italy, Denmark, Ireland, Great Britain, and Greece, and was positively associated with age in Northern Ireland. On the issues of controlling pollution and protecting the environment, there was no association with age because agreement on the importance of doing these things was all but unanimous (Rabier & Inglehart 1980, Rabier, Riffault & Inglehart 1983a, 1983b).

Cultural stereotyping can be a powerful influence. In societies conscious of modernity, the values of older persons, like their skills, are commonly dismissed as obsolete. Young and old alike generally assume that to age is to grow more conservative, more fearful of risk and novelty. But such an assumption, in the words of an American philosopher (Esposito 1987, 219–21),

> requires a more concerted defense than is usually made. True enough, the aging individual is familiar with the events and times of a world gone by, but past events are always fully modern in their own time; to have been part of them is to have been part of a modern era, with all of the excitement and danger that implies. Even if everything is old-fashioned from some later perspective ... everything is not obsolete as well. ... Sometimes the present is feeble and enfeebles the imaginations of those shaped by it, while the past is vigorous and progressive – and so too those shaped by it. ... Specific socioeconomic conditions are likely to determine whether a person becomes more conservative or more radical in old age. The conservatism of youth is well known, as is the radicalism of many an ageing revolutionary.

'Conservatism,' 'progressivism;' 'reactionary,' 'radical' – these are spongy concepts, especially when removed from their specific referents. Were Hitler's youthful storm troopers 'progressive?' Were the aged Townsendites (followers of Francis E. Townsend, an elderly California physician who, in 1933, launched the 'Townsend Plan' – to give every American over age 60 a federal lifetime pension of $150 a month as a cure for the Depression) 'conservative?' And even if we as individuals know what we mean by these terms, can we assume that the 'conservative' on one issue is going to be 'conservative' on others as well, or – more to the matter at hand – that these views, however categorized, are causally related to age? For that matter, is receptivity to change always desirable, and nostalgia and conservatism always undesirable? Would it not be more to the point to ask: Receptivity to what *kinds* of change? Nostalgia and conservatism about *what*? Or, for that matter, *which* 'old ideas' in *which* old houses?

With behavior in general, it is useful to distinguish between a particular group's role when it constitutes a small proportion of the total population and when its share is much larger. In relation, specifically, to politics, the ratio of older to younger voters in a political democracy can be presumed important to the extent that the concerns of the two are different or, more important, in conflict. But, as already noted, while differences and conflicts do exist, there is also much commonality among the interests of the aged and the rest of the society, even if that commonality is not always recognized. Moreover, numbers do not always represent power.

In what has been termed the 'grey peril' hypothesis, it has been suggested that widespread political activism among older persons, in combination with their increasing numbers, will result in (a) resistance to local government taxing and spending on behalf of programs lacking in immediate benefits for older people and (b) increased demands for services that principally benefit the elderly at the expense of younger persons.

Tests of this hypothesis in the U.S.A., where the prominence of local taxation on behalf of schools and certain social services provides an especially good opportunity for such tests, have yielded somewhat mixed results. Generally, they have either failed to support it or, where they have seemed to support it, the support could be accounted for in terms of special circumstances (see, e.g., Button & Rosenbaum 1989, Rosenbaum & Button 1989). Where the aged have appeared to be less generous concerning transfers to younger lower-income families, the association has been more with income than with age itself; and where more generous transfers to the elderly have been supported, it has been through the willingness of people in all age groups and not just the oldest. In fact, when it comes to awarding benefits to poor elderly households, one study found that the 'coalition' favoring generous transfers to the elderly appeared to 'consist of a minority of the most supportive elderly combined with the implicit support of the majority of the non-elderly' (Ponza *et al.* 1988, 463).

One point to note about this is the possible operation of what has been termed 'the law of the many' (and its corollary, 'the law of the few'), namely, that the greater the size of the group (above some critical and, depending on the circumstances, varying, mass), the smaller will be its political influence. The main reasons for this seeming paradox are: (a) the greater difficulty a larger group faces in maintaining a united political front – partly because greater size affords greater opportunity for the development of a diversity of viewpoints (and even conflict) within a group, and partly because greater size can encourage the individual members of a group to think that they need not, themselves, do anything toward maintaining political influence because there are so many others available to perform such chores; and (b) the fact that the larger a group, the larger will be the total cost of any additional benefits granted it by the society. It is easier to make special provision – even fairly high per capita provision – for a group that is small than for one that is large. McKenzie (1991) notes, for example, that in the U.S.A., welfare benefits for the aged predictably became just another tax target when, in the 1980s, their total cost reached more than $200 billion. That younger taxpayers were, at the same time, beset with rising Social Security and Medicare tax rates only assisted the process. Congress responded to this development by, among other things: increasing insurance premiums and restricting medical benefits for the aged, reducing compensation for changes in their cost of living, introducing taxation on old age benefits, and increasing the age at which full retirement benefits can be received.

'President Bush's proposed $23 billion additional cut in Medicare benefits', McKenzie observes, is only 'the most recent sign of the elderly's languishing political clout.'

The other point to note about the 'grey peril' hypothesis relates to the commonality of interests between the aged and the rest of the society.

Studies in both the U.S.A. and Europe (Baum & Baum 1980: 78–85, Phillipson 1982, 129–31) have found differences in political views between the aged and the rest of the society (at least through the 1970s) to be slight, inconsistent, and unstable. Apart from a higher endorsement of medical care issues in the U.S.A., there is little evidence of an 'aging vote' or of any trend toward one. We should not be surprised by any of this. People occupy a variety of social positions, by no means all of which are age-related. They also have a variety of values (not to mention information) upon which to base a political judgment. While most individuals tend to show consistency between their social positions and values, there is usually enough normative leeway to permit behaviors that appear to be inconsistent. Moreover, in the fact of role conflict lies a further stimulus to seemingly inconsistent behavior. At any one time, each individual simultaneously occupies a considerable variety of social positions – wife, mother, daughter, friend, worker, Catholic – each with a set of associated behavioral prescriptions and proscriptions, each of which is likely to incorporate some logical inconsistencies within itself and in some measure to be at variance with the blueprints for behavior associated with the others. Because the positions occupied will vary over time and with age, there is the further possibility of role conflict arising out of a lingering adherence to the norms associated with some previous position.

Given the variety of one's positions, both past and present, and the possibility of role conflict arising out of them, it is hardly surprising that, however much one might expect to find a close association between age and interest, there will always be some uncertainty as to what proportion of a particular age group will actually hold to this interest strongly enough to act on it. Moreover, because old people have had longer to occupy these different positions and to play the roles associated with them, and also because they would be members of cohorts that would, by definition, have experienced the greatest degrees of social change, their behavior could be expected to be particularly affected by any tendency to adhere to the norms associated with some prior social position. No more in political behavior than in anything else should we expect older people to be particularly homogeneous.

Then, too, there is always the more general issue of social policy; of tradition and which of a people's values and personality traits their leaders appeal to – whether, so to speak, they appeal to the best or to the worst in them. A comparison of provision for children in

France and the U.S.A. is instructive. An American delegation studying childcare in France concluded that the country, although at a level of economic development similar to that of the U.S.A., was 'far ahead ... in insuring that its young children are well and safely cared for.' The French system is a blend of childcare, education and health services based on free full-day preschool programs, subsidized day-care centers, and licensed care in private homes for infants and toddlers. The noncompulsory preschool programs, which serve nearly 90 per cent of French three- to five-year-olds, offer language, arts, exercise, crafts, and play. According to the report, the French system also features 'intensive training and fair compensation for preschool teachers and others who take care of younger children, a free preventive health program for all young children, and attention to the architecture and safety of day-care centers.' As one of the delegation's members put it, 'Coming from a country [that is, the U.S.A.] mired in turmoil over child care, it was striking to see in France a shared consensus about the importance of children and the willingness to put the necessary financial resources behind it' (Lawson 1989). For present purposes, it is important to note that the French system is largely financed by tax revenue and therefore subject to approval or rejection by the voters: this in a country where, in contrast with the U.S.A., those 65+ years of age are some 6 per cent, and those 75+ years of age some 37 per cent, higher a proportion of the voting age population than are their counterparts in the U.S.A. In Sweden, another country noted for its children's programs, these proportions are, respectively, 36 and 46 per cent higher than the U.S.A.'s. It would seem that rather more than age distribution is at work here.

Nevertheless, there is growing evidence that, so far as tax-funded social supports are concerned, the position of the elderly in a number of industrialized countries is improving at a time when that of younger persons is deteriorating (Preston 1984; Thomson 1989). Some of this discrepancy could well be due to age differences in political power and attitude. But it seems likely that most of it results, instead, from recent (and presumably only temporary) changes in political ideology and leadership, together with the fact that pension programs (which are the main element in tax-funded supports for the aged), as contrasted with schooling, housing, income maintenance, and maternal and childcare (which are the main elements of tax-funded support for the rest of the society) tend to be funded in ways less susceptible to either sudden legislative change or administrative tinkering. The generation who reached adulthood during and immediately after World War II, in receiving substantially more from their respective countries' welfare systems than they put in (Thomson 1989), may well prove to have been uniquely favored. They were born late enough to benefit from programs initiated too late for most of their predecessors, and early enough to avoid being directly affected by the 'piecemeal erosion' of these programs that their successors are being (or, it is predicted, will be) subjected to because of the (hardly surprising) failure to eventuate of the indefinite continuation of economic and demographic growth on which the institution of these supports was largely predicated. But the achievement of greater demographic stability following upon the passage through the system of the distortions consisting of the swollen 'baby boom' generation and its constricted parental generation – plus a renewed appreciation of the necessity of making social, and not merely individual, provision for the satisfaction of human needs – should be enough to correct any imbalance for the generations that follow. So far as specifically demographic factors are concerned, any age imbalance in tax-funded welfare support will arise from conditions that are likely to be only temporary. The more important factors in the equation would appear to be political ideology and leadership, and here the force of demographic conditions, as such, is negligible.

REFERENCES

Baum, M. & R. C. Baum 1980. *Growing old: a societal perspective.* Englewood Cliffs, N.J.: Prentice-Hall.

Blake, J. 1971. Abortion and public opinion: the 1960–1970 decade. *Science*, Feb. 12.

Boulding, K. 1966. The economics of the coming spaceship earth. In *Environmental quality in a growing economy*, Henry Jarrett (ed.). Baltimore, Md: Johns Hopkins University Press.

Button, J. W. & W. A. Rosenbaum 1989. Seeing gray: school bond issues and the aging in Florida. *Research on Aging* 11(2), 158–73.

Daly, H. E. 1971. Toward a stationary-state economy. In *Patient Earth*, John Harte & Robert H. Socolow (eds.). New York: Holt, Rinehart & Winston.

Daly, H. E. 1977. *Steady-state economics.* San Francisco: W. H. Freeman.

Day, A. T. 1991. *Remarkable survivors: insights about successful aging among women.* Washington, DC: Urban Institute Press.

Day, L. H. 1971. Concerning the optimum level of population. In *Is there an optimum level of population?*, S. Fred Singer (ed.). New York: McGraw-Hill.

Day, L. H. 1978. What will a Z.P.G. society be like? *Population Bulletin* **33**(3), Population Reference Bureau, Washington, DC.

Esposito, J. 1987. *The obsolete self: philosophical dimensions of aging.* Berkeley: University of California Press.

Hamblin, C. L. 1970. *Fallacies.* London: Methuen.

I.L.O. (International Labour Office) 1986. *Year Book of labour statistics.*

Joske, S. 1989. *The economics of migration: who benefits?* Background paper from Legislative Research Service, Parliament of the Commonwealth of Australia, Canberra.

Kelley, A. C. 1972. Demographic changes and American economic development: past, present, and future. In Morss & Reed (eds).

Kelley, A. C. 1989. Economic consequences of population change in the third world. *Journal of Economic Literature* **26**(4), 1685–1728.

Krosnick, J. A. & D. F. Alwin 1989. Aging and susceptibility to attitude change. *Journal of Personality and Social Psychology*, **57**(3), 416–25.

Lawson, C. 1989. France seen as far ahead in providing child care. *New York Times*, November 9.

McKenzie, R. B. 1991. The retreat of the elderly welfare state. *Wall Street Journal*, March 12.

McKibben, B. 1989. *The end of nature.* New York: Random House.

Mishan, E. J. 1967. *The costs of economic growth.* London: Staples Press.

Morss, E. R. & R. H. Reed (eds) 1972. *Economic aspects of population change*, Vol. 2 of Commission *Research Reports*. U.S. Commission on Population Growth and the American Future. Washington, DC: U.S. Government Printing Office.

Nielsen, J. T. 1989. Immigration and the low-cost housing crisis: the Los Angeles area's experience. *Population and Environment* **11**(2), 123–40.

Phillipson, C. 1982. *Capitalism and the construction of old age.* London: Macmillan.

Ponza, M., M. Corcoran, G. J. Duncan & F. Groskind. 1988. The guns of autumn? Age differences in support for income transfers to the young and old. *Public Opinion Quarterly* **52**(4), 441–66.

Preston, S. H. 1984. Children and the elderly: divergent paths for America's dependents. *Demography* **21**(4), 435–57.

Pryor, R. J. 1977. The migrant to the city in South-east Asia – can, and should we generalise? *Asian Profile* **5**(1), 63–89.

Rabier, J.-R. & R. Inglehart 1980. *Euro-barometer 10: national priorities and the institutions of Europe, November 1978* (machine-readable data file). 1st I.C.P.S.R. edn. Ann Arbor, Mich.: Inter-university Consortium for Political and Social Research.

Rabier, J.-R., H. Riffault, & R. Inglehart 1983a. *Euro-barometer 16: noise and other social problems, October 1981* (machine-readable data file). 1st I.C.P.S.R. edn. Ann Arbor, Mich.: Inter-university Consortium for Political and Social Research.

Rabier, J.-R., H. Riffault & R. Inglehart 1983b. *Euro-barometer 17: energy and the future, April 1982* (machine-readable data file). 1st I.C.P.S.R. edn. Ann Arbor, Mich.: Inter-university Consortium for Political and Social Research.

Ringen, S. 1987. *The possibility of politics: a study in the political economy of the welfare state*. Oxford: Oxford University Press.

Robinson, W. C. 1972. comment. In Morss & Reed (eds).

Rosenbaum, W. A. & J. W. Button 1989. Is there a grey peril?: retirement politics in Florida. *The Gerontologist* **29**(3), 300–6.

Schumacher, E. F. 1973. *Small is beautiful*. London: Blond & Briggs.

Smil, V. 1990. Planetary warming: realities and responses. *Population and Development Review* **16**(1), 1–29.

Spengler, J. J. 1972. Declining population growth: economic effects. In Morss & Reed (eds).

Thomson, D. 1989. The welfare state and generation conflict: winners and losers. In *Workers versus pensioners: intergenerational justice in an aging world*. Paul Johnson, Christoph Conrad & David Thomson (eds.). Manchester: Manchester University Press.

U.N. (United Nations) 1985. *Demographic yearbook, 1985*. New York.

U.N. 1989. *World population prospects 1988*. New York.

Wander, H. 1977. Short, medium and long term implications of a stationary or declining population on education, labour force, housing needs, social security and economic development. In International Union for the Scientific Study of Population, *International Population Conference: Mexico 1977*, Vol. 3, 95–111.

World Commission on Environment and Development (The Brundtland Report) 1987. *Our common future*. Oxford: Oxford University Press.

Zapf, W. 1984. Welfare production: public versus private. *Social Indicators Research* **11**(3), 263–74.

5

POLICY ALTERNATIVES: DEMOGRAPHIC

INTRODUCTION

Before considering the policy alternatives to be undertaken in these populations in response to the expectation of older age structures and numerical declines, let us, firstly, put the matter into some perspective. Much of the present concern about the 'burden' of the aged seems to have its origin less in knowledge than in ignorance about the characteristics and needs of the elderly and the nature and extent of their social contribution (whether past or present). It also doubtless owes something to the relative novelty of having to rely so much on social, as contrasted with private, sources for provision of the services and facilities deemed necessary to an older population; for costs are more readily assessed and more often converted into monetary terms available for public consumption when the responsibility for such provision falls upon society as a whole than when it falls upon the more restricted entity of recipients and their families. And, of course, any concern over costs is but heightened by the fact that the increase in the demand for these services and facilities seems to have arisen so suddenly – largely, it must be remembered, in response to changes in program, not age structure.

Yet, the aged, as noted earlier, are not only largely responsible for most of their own care and that of their spouses, but can – and already do – perform a number of services to both kin and the larger society. That these services are not ordinarily distributed through the marketplace detracts not at all from their importance.

Moreover, the very fact of having a smaller proportion in the younger ages should make possible at least some financial savings, so far as society in general is concerned. Whether these will be enough to offset increased expenditures occasioned by the correspondingly higher proportions in the older ages is essentially a matter of definition and time-perspective. Children are particularly expensive in industrialized societies. A costing of goods and services in such a society that was complete enough to include those distributed outside the marketplace as well as through it would doubtless show the financial cost of children to be higher than that of the aged. The economic dependency of children in such societies – for food, clothing, and shelter; for supervision; for health care; for education and training – is essentially total. Unlike that of their counterparts in nonindustrialized societies, the economic contribution of children in industrialized societies is virtually nil – and declining still further with the advent of high-technology agriculture and the passing of the family shop.

The aged, on the other hand, in addition to the goods and services they might currently produce, will have had a lifetime of production behind them. In those of these countries in which programs for the aged were late in getting started, there will be a period during which current outlays for these programs are likely to exceed the current revenues earmarked for this purpose. But the fact that future generations can be expected to reach old age after long periods of participation in the work force holds out the likelihood that, before long, most old people in these countries will be financially supported directly out of funds they have paid in during their years of peak earning. Achieving such a goal requires little more than the application of appropriate taxation, social security, and similar forced savings programs of a sort that most of these countries already have well in place.

On another plane is the matter of the coin in which program costs are measured. If the costs of programs for the aged (or of any social programs, for that matter) could be expressed, not in pounds or dollars, but in something like 'missile-' or 'bomber-equivalents,' instead, it might go a long way toward allaying the concern that the costs of social programs seem to engender. Such a salutary practice would not require much additional effort. That it might eventually be followed on occasion is suggested by the way one of the U.S.A.'s leading newspapers not long ago reported the crash of a third B-1 bomber within a period of 14 months. The $840 million cost of these three bombers was compared with the $375 million cost of three 747 jetliners, the $964 million fiscal 1988 budget for Fairfax County, Virginia (a sprawling suburban area outside Washington, DC, with a population of more than half a million), and the $1,100 million for fiscal 1987 of Head Start (a nationwide program established to improve the health, learning capacities, and education of children in the most disadvantaged sectors of the population) *Washington Post* (November 19, 1988).

But for present purposes, let us assume that these concerns, however unrealistic or overstated they may in some instances be, are nonetheless genuine and deserving of our attention, and then proceed from there to a consideration of what policy measures are called for, commencing with the specifically demographic.

CHANGING THE DEMOGRAPHIC SITUATION

One way to address the consequences of numerical declines and older age structures would be to change these presumably undesirable demographic conditions themselves. The goal would be a younger age structure and a population growth rate no less than zero. The demographic means to such a goal are three: (1) increase fertility, (2) increase net immigration, and (3) allow mortality to increase at the older ages.

Increasing fertility

The approaches to a higher fertility level are also three: (1) force people to bear children they do not want, (2) increase the desire (or, at least, the willingness) to have greater numbers of children, and (3) remove obstacles to the bearing and rearing of children so it will be easier for people to have more.

Forcing increased childbearing upon people (by limiting access to contraceptives and abortion) is, at the least, ethically – and, in a democratic society, politically – questionable. It is also unlikely to work – at least more than very temporarily – in a society that has already brought fertility under extensive control. The experience of Ceausescu's Romania illustrates the point. At a time when considerable control was already being exercised over fertility (the crude birthrate fluctuated among the mid-20s for the first ten years after World War II), the government, in September 1957, decreed that abortion would be available on request. In the years that followed, abortion rapidly became the major means of birth control. By 1966, the birthrate had dropped to 14.3 per thousand (from 24.2 in 1956, the last full year before the decree) (U.N. 1975, Table 21), and the Total Fertility Rate to 1.88 (from 2.89 in 1956) (calculated from data in U.N. 1979, Table 6). Simultaneously, abortions soared, rising from 30 per 100 live births in 1958 to 408 in 1965 (Tietze & Murstein 1975, and Tietze 1977, cited in Berelson 1979, 209). Perceiving this fertility decline as a threat to its development plan, the Romanian government, in October 1966, suddenly issued another decree, this time making abortion legally available only under certain limited conditions: danger to life, risk of a deformed fetus, rape, mother's age over 45, and certain specified life situations (Berelson 1979, 209). The next year saw the birthrate almost double to reach 27.4 (it actually reached 39 between the ninth and eleventh months following the decree (Berelson 1979, 209–10)), and the Total Fertility Rate also almost double to reach 3.66 (calculated from data in U.N. 1979, Table 6). Yet, within four years, the birth rate had declined to 21.1 and the Total Fertility Rate to 2.89 (Berelson 1979, 210, and U.N. 1975, Table 21); and by 1985, the birthrate and Total Fertility Rate were both down to almost their pre-restriction levels: 14.8 in the case of the former and 2.05 in the case of the latter (World Bank 1986). To be sure, the prohibition resulted in more births than there would otherwise have been, but only at a considerable cost: the frustration of parental choice, increased maternal mortality in consequence of illegal abortions, and present and future dislocation in education, housing, employment and, later, pensions, as this swollen cohort reaches successive age levels (H. R. Jones 1981, 186). Forcing people to bear children they do not want can – and should – be dismissed out of hand.

No such limitations are involved in increasing the number of children people want (or are willing) to have. But a policy to this end would imply both a knowledge of reproductive motivation and the ability to manipulate the determinants of that motivation, and nearly half a century's research has done little more than narrow the range of conjecture about reproductive motivation. We still cannot account, really, for why one couple has, for instance, two children and another four; or, at the more general level, for cohort differences in the distribution of family sizes or the timing of births. The causes of different patterns of childbearing, especially where, as in the countries under consideration here, fertility is subject to a large measure of direct control, remain very largely the conundrum they have always been.

In what seems like something of a response to the frustration this has engendered, demographers are increasingly focusing fertility research on what they term 'proximate determinants:' the proportions in sexual unions and the age at entry into such unions, the proportions using contraception, and ages at first and last birth, for instance. This adds little to our understanding, however, for it only moves the issue to another plane of inquiry: why these particular ages at entry? these ages at birth? these proportions bearing these numbers of children?

Most of what is, in fact, known about reproductive behavior is at a much higher level of generality (L. H. Day 1977 and 1985). What can be said, in general, about the causes of a particular pattern of childbearing is essentially what can be said, in general, about the causes of *any* social behavior: (a) a variety of factors will have been at work, some supporting the behavior in question, others impeding it; (b) these factors will have consisted of a variety of elements in (i) the physical environment (including the individual actor's physiological attributes), (ii) the social environment (consisting of institutional structures and norms, as well as of other human beings with whom the actor interacts – both directly and indirectly), and (iii) the personality structures (that is, the respective sums of their psychic predispositions to act) of both the individual actor and those with whom he or she interacts; (c) the particular factors involved – and the significance of each – will have been different for different actors; different at different periods of time (different, for example, in times of war than in times of peace, in periods of relatively full employment than in periods of widespread unemployment); and different at different stages in the actor's lifetime (different for 0-parity as against 1-parity women, for example, for 25-year-olds as against 35-year-olds, for mothers whose youngest child is still at home as against mothers whose youngest is in school all day); (d) related to this is the fact that, in any specific behavioral situation, there is seldom, if ever, only one possible behavior. Instead, there is usually a whole range of possibilities; the particular one followed in any specific instance being the result of an interacting combination of (i) various elements in the environment (both physical and social), (ii) the personality structures of the actors involved, (iii) these actors' 'definitions of the situation,' and, finally, (iv) chance; (e) conscious choice between alternative behaviors in any behavioral situation is a matter of degree, ranging between the one extreme of a careful weighing of the costs and benefits attributable to a given course of action and the other extreme of a virtually complete absence of any such consideration; (f) human behavior is a process taking place in time, with what happens at a prior stage or period having a bearing on what happens at a later one; (g) a given condition (like, for example, the cessation of childbearing after two children) can have a variety of different origins; (h) similarly, a particular causal element can give rise to a variety of consequences, depending on what other elements it is mixed with in the causal brew.

Specifically with respect to fertility, (i) a given pattern will necessarily have been the result of interaction: interaction between persons of the opposite sex, of course, but also – both directly and indirectly – between potential parents and what is likely to have been a considerable variety of others; others ranging from close relatives and friends possibly all the way to the personalities presented through the media of mass communication; and finally, and again with respect specifically to fertility, (j) whatever the factors in the causal equation, they must inevitably have worked through the intermediate variables of intercourse, conception, gestation and parturition. They must, that is, have affected the timing and frequency of intercourse, the likelihood of conception, and the likelihood that a product of conception would survive long enough to eventuate in a live birth (Davis & Blake 1956).

These are elemental points, generally applicable (with the exception of the last two) to all social behaviors. But they are at a very high level of generality. They provide an overall framework for understanding social behavior, but they tell us nothing about the behavior of particular individuals in particular circumstances: nothing as to either the identities of the various causal elements at work, or the relative degrees of their significance.

Moreover, contrary to the assumptions underlying the way it is commonly treated in fertility analyses – particularly, but not exclusively, by those who come to such analyses by way of a training in economics – decision-making about childbearing – as about so much else – is not invariably marked by a high degree of consciousness. It is one thing to note that people make conscious choices, and that they make them within a broad normative framework that establishes limits and guidelines respecting both the goals to be sought and the means appropriate to the attainment of those goals; it is quite another to declare that conscious choosing is the characteristic form of human behavior or, more germane to present considerations, that it is the characteristic form with respect to the frequency and timing of childbearing. Most human behavior – fertility or otherwise – entails little conscious decision-making among well-defined alternatives. In any particular instance, the individual actor is ordinarily but dimly aware of why he or she is behaving in one way rather than another. Human beings may be calculating and reflective animals, but we need hardly conclude from this that all – or even most of some of the most socially significant – of their behavior is truly cognitive. We need, in short, to recognize the supplementary existence of behavioral 'drift;' of behavior moving by small degrees along the continuum of possibilities. We need, that is, to recognize that human behavior has to it an on-going, developmental, 'incremental' side – a side in which interaction with others plays a particularly significant role – and that most of the countless 'decisions' by which individuals influence their own lives and those of others are made all but imperceptively and with relative ease. Decision-making is, in short, neither as comprehensive nor as temporally consistent as most of the theories, particularly those of economics, would have us believe.

So far as childbearing is concerned, the nature of both the data and the variables used by the researchers has permitted only a knowledge of general probability and direction. But, given that the goal would be the generality of behavior rather than the behavior of specific individuals, a knowledge of general probability and direction should be enough to act on when it comes to establishing effective policy. Why, then, have pronatalist policies been so unsuccessful (see, e.g., Höhn 1987, Monnier 1989)? Demeny (1987) asserts that governments, being ill-equipped 'to engineer and guide appropriate value changes,' are confined essentially to the manipulation of 'economic levers,' and that, so far, they have not manipulated the most effective of these. Accordingly, he proposes that, if they are to increase fertility, governments must re-establish the material link that once presumably existed between fertility and old-age security by allocating a specified fraction of an individual's social security contributions directly to his or her parents.

Although one of the more sensible, sophisticated, and easily administered of the genre, this particular pronatalist policy proposal ignores the fact that many people in low-fertility societies already render economic help to their aged parents. Moreover, in starting with its premise about the limitations on governmental action, it fails to consider the possibility that governments may not be quite so confined to purely economic levers, after all. Surely governmental policy on, for example, education, foreign affairs, or environmental issues could also be expected to have some determinative effect on whether the social setting was more or less conducive to childbearing.

Apart from the question of whether it would actually work, probably the main drawback of this proposal is that, contrary to its author's claim, it would not be very fair. Children do entail economic and emotional costs for their parents, but they also hold out the possibility of substantial psychic benefits which, like the proposed extra income from children's social security payments, are unavailable to the childless. Nor do parents bear all the costs of raising their children. The whole society shares in this: in, for example, the economic costs of providing schooling and other social services; the emotional costs occasioned by children's intrusion upon adult privacy, space, and the enjoyment of life; and the costs from higher crime rates associated with youth.

Because parents do experience special needs, removing obstacles to childbearing – like inadequacies in childcare, maternity leave, medical care, or housing – would more likely meet the criterion of equity. But most of the countries under consideration already provide some such assistance – without any noticeable boost to fertility in consequence. In this way they are encountering much the same general lack of demographic response as met the avowedly pronatalist policies some of their governments enacted during the 1930s (Glass 1940, Chs 4 and 5). Of these policies, only those of Nazi Germany appear to have had any appreciable pronatalist result; but whether the increased German birthrate at the time indicated an actual change in fertility or merely a change in the timing of the births that would have occurred irrespective of the policy is something the social disruption of World War II, which commenced soon afterwards, prevents us from ever knowing (Glass 1940, Ch. 6). Expanding and extending assistance to parents could be supported on grounds of social justice, but it is doubtful whether the effect of such an undertaking on fertility would be more than marginal.

Increasing immigration

Two characteristics of immigration hold particular appeal for host societies: (a) its apparent amenability to precise regulation as to rate, direction, and composition, and (b) the possibility it holds out of gaining numbers of people and work skills without having to bear the prior costs of nurture and training. Neither of these is in any way a certainty. Moreover, while the social and economic costs of international migration – in both host and donor populations – tend to be spread among the many, the flow of benefits tends to be highly concentrated. In host populations, benefits tend to flow mostly to building contractors, developers, ethnic leaders, and, of course, to some (although hardly all) of the immigrants themselves; in donor populations, they tend to flow primarily to those of their kin to whom the migrants send remittances.

Immigration entails problems – for the migrant, the host society, even the donor society; the types and seriousness of which are affected by the size and rate of immigration, as well as by whether the immigrant wants to remain in the host country and does so, and the extent to which he or she visits and sends remittances to the country of origin. The consequences of immigration can flow both ways. If there are some on whom it bestows benefits (usually, although by no means exclusively, of a narrowly economic nature), there are others (including those who may otherwise be benefiting) from whom it exacts a cost. Sometimes this cost is economic (the loss of the migrants' labor, skill, capital, or enterprise to those who remain in the place of origin, for example; or, in the host country, the threat migrants pose to employment levels, wage rates, or working conditions, or the additional competition – with consequent inflationary pressures and declines in ranges of choice – migrants introduce into the housing market). More often this cost is emotional, social, cultural, even environmental: some of it borne by those in the host society, some by those in the donor society (see, e.g., Bohning 1975, Bell 1979, Cohen & Lewis 1979, Van de Ven & Diphoorn 1981), some by the migrants themselves (see, e.g. Cronin 1970, Içduygu 1990). As far as understanding the full range of its implications is concerned, it is unfortunate that research on international migration is currently so much in the hands of economists; for the tools and perspectives of economics are simply not suited to consideration of the likes of, for example, ethnic and religious animosity and prejudice, intergenerational conflict arising out of cultural differences within families, scapegoating, homesickness, and social marginality; or ethnic and religious bloc politics, institutional change, culture and role conflict: all aspects of international migration hardly less real or important for the short shrift researchers commonly accord them than the economic aspects on which researchers lavish so much of their attention, instead.

So far as population aging and decline are concerned, immigration is, at best, a merely temporary expedient. Over time, the inevitable concentration of immigrants within certain age groupings can markedly distort an entire age structure, the extent of this distortion depending on the immigrants' numbers, fertility levels, and durations of residence. Recent calculations for Australia (Young 1989), a country in which immigration is widely supported as a solution to population aging, illustrate the point. In comparison with what would be produced in the absence of immigration, continuation of that country's present high rate of annual net immigration (approximately 0.9 per cent per capita – the highest in the Western World, incidentally), in combination with the continuation of overall fertility at slightly less than replacement level, would produce a more distorted age structure and an only slightly lower proportion in the oldest ages. Accepting some immigration into Australia as inevitable, Young makes two additional sets of illustrative calculations: one on the basis of an annual net immigration level of 150,000 and the other on one of 50,000. In historical terms, even the lower of these is quite high. In combination with the continuation of fertility at the somewhat below-replacement level of 1987, the higher migrant intake would, indeed, yield a slightly smaller proportion aged 65 and over. By mid-century, however, the elderly proportion of the population would have stabilized within the 20–22 per cent range, regardless of which of these two immigration levels had been in effect; and, more important: while the proportions of elderly would be essentially the same, the differences in population size would be substantial. In the 60 years between 1986 and 2046, the higher immigration level would have resulted in a population increase of some 90 per cent (raising it from 16 to 30 million), as against an increase of some 38 per cent (from 16 to 22 million) with the lower rate of immigrant intake. A slight retardation in the aging of the population (a difference of less than 2 percentage points between an elderly proportion of approximately 19.5 per cent versus one of some 21.4 per cent) would have been achieved by means of a very high level of immigration (with all its attendant difficulties) and at the cost – in living conditions, the environment, and so on – of the eight million more persons than would have been added by what, historically, would have been the still high annual intake of 50,000 immigrants. With no immigration at all – presumably a political impossibility in Australia at the present time – the same set of fertility and mortality assumptions would result, by the year 2046, in 23.0 per cent aged 65 and over and total numbers at the arguably more manageable level of 18 million, instead (Young 1990, 16–17).

Nor can immigration be depended upon to introduce a lasting element of higher fertility into a population that has ceased to replace itself. Higher-fertility immigrants to low-fertility countries seem soon to become quite as receptive as the natives to these countries' inducements to low fertility. (L. H. Day 1983, Tribalat 1987, 370). Moreover, the usual difficulties of adjustment and adaptation associated with migration would be aggravated by the fact that higher-fertility immigrants would necessarily tend to come from populations with markedly different cultural and, possibly, racial characteristics. Even in generally favorable circumstances, the immigrant is apt to be something of a second-class citizen (if citizen at all) – economically, socially, politically – especially if he or she is one of a large influx and has been recruited (as is likely in the countries under consideration) for the less desirable employments. Any marked differences in culture or race merely reinforce this tendency. Second class citizenship is no buttress for a democracy, nor can its existence be considered desirable – for either immigrant or native.

Overall, the immigration 'solution' stresses too much the short-run and economic – especially labor-force – aspects of the demographic changes attending low fertility, and too little the long-run, noneconomic, social and personal aspects. In the process, it tends to pass over the basic humanity of everyone concerned: immigrant, native or member of the immigrant's network in the place of origin. To rely on immigration to resolve the difficulties expected from low fertility is seriously to risk a society's future wellbeing and identity on behalf of a policy whose present efficacy can, in the best of circumstances, be little more than questionable.

It is also to overlook readier – and less costly – solutions closer to home. The size and composition of the work force is a case in point. As long as a population contains people who, for whatever reason, are not in the work force but who could – with the adjustment of work schedules, the provision of day-care facilities for young children, or the further development of education and training programs, for example – be rendered willing and able to enter it, there is at least the potential for solving a presumed labor shortage by recruitment of persons already at hand. This might require some effort, even some expenditure. But the cost to the society as a whole of assisting or reimbursing individual employers for the additional effort or cost entailed – or of no longer permitting employers to socialize the costs of some of their operations by importing workers from abroad – could hardly be as great as those entailed in attempting to resolve labor shortages through immigration, instead. Some indication of the potential base for such an approach can be found in international differences in relative labor force participation rates among men and women. In 1980, for instance, while the female participation rate at age 35–39 was 85 per cent as high as the male in Sweden, it was only 61 per cent as high in Australia. At age 40–44, the corresponding figures were, respectively, 89 and 65 per cent (calculated from data in U.N. 1984, Table 26).

Increasing mortality

Increasing mortality at the upper ages would add to numerical decline, but the costs thus entailed would presumably be more than offset by the gains associated with the more youthful age structure that would result. Two procedures are possible: (a) direct intervention through suicide and euthanasia and (b) indirect intervention through the withholding of life support.

One's view of the morality of such action depends on one's view of people's rights over their bodies and the importance of each individual having as long a life as possible, irrespective of its quality. While medical progress may have led to something of an 'expansionary vision of health' incorporating the belief that the 'seemingly inherent finitude of the body' can be overcome (Callahan 1987, 18–19), there is enough clinical evidence of the wide practice – under medical auspices – of both suicide and euthanasia in these societies to suggest that the opposite view is also a basis for action. Moreover, in the growing number of 'living wills' being written in an attempt to ensure against being victimized by the undue application of medical technology and ministration, and the interest expressed by many adults in acquiring the knowledge and means by which to ensure they have a peaceful death (Bates 1980), there is evidence that suicide and euthanasia would be more widely practiced if circumstances permitted. Apparently, there are many who fear death less than they fear dying in a painful or long drawn-out way; less than they fear becoming senile caricatures of themselves, 'turning into grandma who keeps wandering away, who cannot control her bladder and her bowels, who spills her food, who repeats herself all the time, and is treated like a baby' (Bates 1980).

Certainly there are gains to be had in honoring those requests to be 'gently eased out of their pain and life' that come from stricken patients 'whose days and nights,' in the words of a well-known philosopher speaking from personal experience, 'are spent on mattress graves of pain' (Hook 1987). But whether further relaxation of the strictures on suicide and euthanasia in these societies would have much effect on their age structures is another matter. Most old – even very old – people are capable of looking after themselves (sometimes with a bit of help, of course). No rational social policy could be aimed at ridding the society of these. Such a policy would have to be restricted to the most helpless and those suffering extreme pain. How many would this be? The Royal College of Physicians (1986) has estimated that approximately 10 per cent of the population of England and Wales is physically disabled (excluding sensory and mental disorders), with about 10 per cent of the disabled population being under the age of 45, 30 per cent between 45 and 64, and 60 per cent age 65 or older. The College estimates, further, that some 20–30 per cent of these disabled persons (that is, some 2–3 per cent of the total population) are 'severely or very severely' disabled.

To obtain an idea of the maximum reduction in aging that might be achieved through permitting greater access to suicide and voluntary euthanasia, let us assume that the age-distribution of the 'severely or very severely' disabled is the same as that of the disabled population as a whole, and that suicide or voluntary euthanasia is, in fact, applied to *all* of that 2–3 per cent classified as severely or very severely disabled. Right away (that is, on the basis of the 1985 population structure), the population of the United Kingdom would become younger: the median age would drop by 0.96 of a year on the 2 per cent assumption and by 1.47 years on the 3 per cent assumption, while the proportion age 65 or over would decline by 1.20 percentage points (from 15.16 to 13.96 per cent) on the 2 per cent assumption and by 1.38 percentage points (from 15.16 to 13.78 per cent) on the 3 per cent assumption. Applying the same approach to the stable (but much more aged) population achieved in the year 2160 through continuation of 1985 fertility levels also results in a younger age structure: a median age 1.16 years lower on the 2 per cent assumption and 1.76 years lower on the 3 per cent assumption, and a proportion 65 or over that was 0.73 percentage points lower (23.54 per cent down from 24.27 per cent) on the 2 per cent assumption and 1.11 percentage points lower (23.16 per cent down from 24.27 per cent) on the 3 per cent assumption. It must be emphasized that these are calculations based on extreme assumptions. They do show a more youthful age structure – but not by much. There would be undoubted benefits in having a more liberal access to suicide and euthanasia, but they would be in the form of reduced economic and psychic costs attending terminal care. Demographic gains would be minimal.

CONCLUSION

A specifically demographic 'solution' to the undesirable conditions some see as emanating from older age structures and smaller population sizes would necessarily entail the employment of means that are either morally objectionable, socially undesirable, of too-limited efficacy, or more than a little likely to produce conditions of even greater undesirability than those they were intended to counteract. It is social, not demographic, policy that offers the possibility of a solution to whatever undesirable consequences might flow from the expected demographic changes. This matter is addressed in Chapter 7.

So far as demographic policy, itself, is concerned, human and social ends would be best served by the following:

(a) With respect to mortality, ensuring as much as possible that people live active, healthy lives – and then, when it comes time for them to depart life, ensuring that they are enabled to do so with a maximum of dignity and a minimum of pain and suffering (both to themselves and to their loved ones);

(b) With respect to immigration, putting the emphasis on making do with the population already in place: encouraging them, training them, providing them with a social and natural environment conducive to their own wellbeing – and, also, to the future wellbeing of the generations to follow; and

(c) With respect to fertility, establishing the goal – after a period of numerical decline – of no more than replacement: attained through maintaining a social and natural environment that was, firstly, favorable to children and did not unduly discourage potential parents about having and rearing children; but which, secondly, would be one in which (i) no unwanted child was born, (ii) the decision to bear or not to bear a child was made solely by the potential parents, and (iii) this decision about childbearing was made in a social and cultural context that defined a three-child family as large (L. H. Day & A. T. Day 1964, 246).

REFERENCES

Bates, E. 1980. Radio talk, Australian Broadcasting Commission program, The body program, February 10.

Bell, R. M. 1979. *Fate and honor, family and village*. Chicago: University of Chicago Press.

Berelson, B. 1979. Romania's 1966 anti-abortion decree: the demographic experience of the first decade. *Population Studies* 33(2), 209–22.

Bohning, W. R. 1975. Some thoughts on emigration from the Mediterranean basin. *International Labour Review* 111(3), 251–77.

Callahan, D. 1987. *Setting limits – medical goals in an aging society*. New York: Simon & Schuster.

Cohen, J. M. & D. B. Lewis 1979. *Local organization, participation and development in the Yemen Arab Republic*. Working Note no. 6, Rural Development Committee, Yemen Research Program. Ithaca, NY: Center for International Studies, Cornell University.

Cronin, C. 1970. *The sting of change: Sicilians in Sicily and Australia*. Chicago: University of Chicago Press.

Davis, K. & J. Blake 1956. Social structure and fertility: an analytical framework. *Economic Development and Cultural Change* 4, 211–33.

Day, L. H. 1977. Models for the causal analysis of differences in fertility: utility, normative, and drift. In *The economic and social supports for high fertility*, Lado T. Ruzicka (ed.). 489–501. Canberra: Department of Demography and Development Studies Centre, Australian National University.

Day, L. H. 1983. *Analysing population trends: differential fertility in a pluralistic society*. London: Croom Helm.

Day, L. H. 1985. Illustrating behavioral principles with examples from demography: the causal analysis of differences in fertility. *Journal for the Theory of Social Behaviour* 15(2), 189–201.

Day, L. H. & A. T. Day 1964. *Too many Americans*. Boston, Mass. Houghton Mifflin.

Demeny, P. 1987. Re-linking fertility behaviour and economic security in old age: a pronatalist reform. *Population and Development Review* 13, 128–32.

Glass, D. V. 1940. *Population policies and movements in Europe*. Oxford: Oxford University Press.

Höhn, C. 1987. Population policies in advanced societies: pronatalist and migration strategies. *European Journal of Population* 3(3/4), 459–81.

Hook, S. 1987. In defense of voluntary euthanasia. *New York Times*, March 1.

Içduygu, A. 1990. *Migrant as a transitional category: Turkish migrants in Melbourne, Australia*. Ph.D. thesis, Department of Demography, Australian National University, Canberra.

118

Jones, H. R. 1981. *A population geography*. London: Harper & Row.

Monnier, A. 1989. Bilan de la politique familiale en république democratique allemand: un reexamen. *Population* **44(2)**, 379–92.

Royal College of Physicians 1986. Physical disability in 1986 and beyond. *Journal of the Royal College of Physicians of London* **20(3)**, 161–94.

Tietze, C. 1977. Induced abortion: 1977 supplement, Induced abortion: 1975 factbook. *Reports on Population/Family Planning* No. 14 (2nd edn.). New York: Population Council.

Tietze, C. & M. C. Murstein 1975. Induced abortion: 1975 factbook. *Reports on Population/Family Planning*, No. 14 (2nd edn.). New York: Population Council.

Tribalat, M. 1987. Evolution de la natalité et de la fécondité des femmes étrangères en R.F.A. *Population* **42(2)**, 370–8.

U.N. (United Nations) 1975, 1984. *Demographic yearbook*. New York.

U.N. 1979. *Demographic yearbook historical supplement*. New York.

Van de Ven, F. P. M. & A. J. Diphoorn 1981. *The impact of emigration on the agricultural development of the Yemen Arab Republic*. Utrecht: Geografisch Institut and Rijksuniversiteit.

Washington Post 1988. Issue of November 19.

World Bank 1986. World population projections. Unpublished, Washington, DC.

Young, C. 1989. Australia's population – a long-term view. *Current Affairs Bulletin* **65(12)**.

Young, C. 1990. Australia's ageing population – policy options. Canberra: Australian Government Printing Service.

6

SOME COMPENSATIONS IN THE TREND TOWARD OLDER AGE STRUCTURES AND NUMERICAL DECLINES

INTRODUCTION

While there are problems associated with older age structures and numerical declines, there are also compensations. For one thing, the alternatives to these demographic developments are assuredly worse than the developments themselves: higher mortality levels to an extent necessary to halt the trend toward older age structures are unacceptable; none of these countries has much to gain, and all have much to lose, from further numerical increases; and, of course, there are limits to how long a population can go on increasing. Moreover, by affording a period of lessened pressure from numerical increase and, at the same time, incorporating a shift in age distribution in a direction that could be expected to result in lower demands on resources – possibly, even, in less support for the economic growth ethic itself – these demographic changes present at least an opportunity to take significant action conducive to a population's living in a more sustainable relationship with its environment.

But discussion of the consequences of these expected demographic developments has largely ignored the possible environmental implications. There is a growing literature relating to living arrangments, the provision of care, and the availability of kin among the elderly. And there has been at least some theorizing about the implications for politics and political behavior, social cohesion (mainly with reference to the consequences of taxing younger generations for the benefit of older ones), and general patterns of behavior. Perhaps because of its alliterative appeal, there is frequent reference to that famous French demographer's surprisingly incorrect assertion (unless old age is assumed as commencing in the mid-30s) that a stationary population (that is, one with a zero growth rate and an older age structure) would be one 'of old people ruminating over old ideas in old houses.'

Probably the most attention has been directed at the economic implications. From a strictly economic standpoint, people are important as producers, consumers, and investors. In addressing the issues raised by older age structures or numerical declines, economists ask about: (a) the labor force (for example, the number of potential participants, whether an older age structure will affect geographic or occupational mobility, and the implications for promotion); (b) the relative costs of youthful and aged dependency; (c) savings and investment (the relative effects on the propensity to save of, for example, household income, the number of one's dependent children, and one's stage in the life cycle); and (d) consumption (prospective changes in, for example, the relative demand for toys versus footwarmers, baby buggies versus false teeth). So far as the economic implications are concerned (and in some contrast with the discussion of noneconomic implications), there has been an abundance of theorizing, but little accumulation of data.

What the approach of the economists emphasizes is *maintenance of a system* – specifically, maintenance of the system of growth capitalism. The goals of economic growth are never inquired into; nor, of course, is the question raised of whether growth capitalism is the most satisfactory means to the attainment of such goals, or even whether it is a means to their attainment at all. This concern about savings and investment, about maintaining economic growth, sidesteps the question of what *kind* of economic growth we might want – or ought, rationally, to have; even of whether continued economic growth is altogether *desirable*. It is not altogether unreasonable to suggest that economic growth actually threatens many of the things most people want and highly value: for example, an unpolluted environment, access to nature, reasonable predictability about the future, social harmony, the avoidance of undue stress, and harmonious relations between generations.

THE OPTIMUM POPULATION

What is at issue here is the concept of the optimum population. Any definition of this optimum implies the existence of individual and social goals the attainment of which is thought to be affected by the number and characteristics of the population. An optimum population is not a goal isolated from other social priorities, nor is it an end in itself. It is a means to the achievement of conditions of life thought desirable.

There are three types of problems to contend with in any consideration of such an optimum: (a) problems of measuring whether demographic characteristics are actually causally related to a particular set of circumstances and, if they are, the extent to which they are, (b) problems of adjustment of cultural differences and change, and (c) problems associated with the diversity of values.

Values lie at the heart of any definition of the optimum. To the problem of measuring the relation between population and particular social conditions, as well as to the difficulty of establishing absolutes in the face of cultural change and differences, must be added the diversity in human values about what ends are to be served by manipulating demographic factors. Because any of them will be to some extent a produce of, and relevant to, a particular cultural setting, both the criteria of an optimum population and the techniques by which to measure movement in the direction of this 'optimum' are always to some degree particularistic and culture-bound.

Most efforts to define the demographic optimum can be faulted on grounds of: (a) the failure to think in terms of a wide variety of human needs, (b) the failure to appreciate the ecological and social limits imposed by our finite planet, and (c) the emphasis on optimum *size* to the exclusion of optimum *characteristics*.

Let us take as our goal for the optimum something general enough to be agreed to by nearly everyone – something like 'happiness' or 'the good life' for all. In determining the bearing of demographic characteristics on such a goal it is necessary to take both the *broad view* and the *long view*. The 'broad' view entails recognizing that human needs are complex and varied, and different at different ages and different stages of life. However, people at all ages and all stages of life need more than food to be fully human. The cultivation of the whole person requires, also, serenity, dignity, order, leisure, peace, beauty, elbow-room – even if human beings can, with their extraordinary powers of adaptability, become, on occasion, inured to severe deprivation respecting one or the other of these needs in much the way they can grow used to chronic pain. A prominent American demographer once remarked,

[T]he disadvantages of a larger population are seen most vividly by those who were born in an earlier era. Often the current inhabitants see nothing wrong with many of the changes that older citizens decry. I feel deprived by the disappearance of open land around Princeton. My children never miss it.

(Coale 1968, 471)

This fluid nature of human values is a major theat to any effort to block progressive deterioration of environmental quality. It underwrites the ever-present tendency to adjust to debasement of the environment by lowering our standards; by, that is, being willing to settle for less.

It is hardly surprising that most discussion of the demographic optimum has been based on narrowly economic criteria of value. Economic conditions are more frequently quantified, and data relating to them are more frequently collected. But the economic criterion is a particularly misleading one because of two implied assumptions contained within it: (a) that economic growth invariably adds to the sum of human happiness and (b) that such growth can be endless – that the limits to economic development are essentially ones of technology and finance, rather than of resources and ecology; rather than, that is, of air, land, minerals, and water, on the one hand, and the ability to maintain essential relationships, on the other (the latter illustrated, for example, by the 'veto of the minority' rule in agriculture which states that all the necessary elements – land, temperature, sunshine, water, seed – must be present in appropriate relation to one another or there will be no crop).

In terms of the broader criteria of human needs, there is no necessary connection between 'the good life' and economic development. In high-consumption societies like those under discussion here, this association can, in many respects, be negative: the greater the economic growth in *general* terms (as illustrated by urban sprawl, highway construction, increased automobile usage, billboards, television advertising, and suburban power mowers, for instance), the greater the deterioration in the quality of life available to the individual – the loss of community, of peace and beauty, of access to outdoor recreation, of clean air and water. The increase in levels of CO_2, the source of the 'greenhouse effect' now coming into being, provides a particularly pervasive (and serious) example. In 1950, world CO_2 emissions amounted to some 1639 times 10^6

metric tons of carbon, and were concentrated in five nations: the U.S.A., the Soviet Union, the United Kingdom, West Germany, and France. Thirty-six years later, in 1986, total emissions were 3.4 times greater (having increased to 5555 times 10^6 metric tons of carbon). Although the U.S.A.'s emissions had nearly doubled in the interim, its share of the total had dropped from the 42 per cent it was in 1950 down to 22 per cent, as per capita emissions from such countries as East Germany, Czechoslovakia, Canada, Australia, and the Soviet Union rose to within 80 per cent of the U.S.A.'s per capita level in the earlier year, and those in such countries as Poland, West Germany, and the United Kingdom rose to within half of it. World economic growth had made the problem truly global (Marland 1989).

It is symptomatic of this topsy-turvy relationship between the 'good life' and economic growth that the environmentally disastrous 1989 *Exxon Valdez* oil spill off the coast of Alaska actually *added* to economic growth – as such growth, despite more than a generation of criticism of the practice, is still commonly (and ludicrously) measured. The time is long overdue to change the purely economic criteria of the good life from those of unlimited expansion to those of conservation, thrift, and careful husbanding of our resources, instead (see, e.g., Boulding 1966).

Population size is not, of course, the only factor determining the quality of life or the relation between a people and its environment. How people experience population size is mediated very largely through their cultural practices and standards. The chances for present and future generations to enjoy 'the good life' depend on the interplay of *population size*, the *rate of consumption* of land, air, water, and minerals, and the *use that is made* of the resources consumed. But while this interplay sets the limits, it ordinarily does so within a wide range of possible variation. We could halt population increase tomorrow and still consume our resources at ruinous rates; and irrespective of the rate at which these resources were being consumed, the uses to which they were being put could eventually result in hardship and chaos.

Taking the *long* view in determining the optimum level of population involves, firstly, recognition that posterity is relevant if for no other reason than that so many of us are going to survive long enough to experience the conditions formed by decisions taken in the present. Even in terms of our own self-interest, we must think well beyond the requirements of the present and immediate future.

A lot of posterity is already here. In Australia, for example, 76 per cent of those alive in 1985 can expect to survive to the year 2015, and 58 per cent of them to the year 2030. Comparable figures for the Netherlands are, respectively 71 and 51 per cent; for France, 70 and 45; and for Italy, 67 and 48.

Moreover, population is not like water issuing from a faucet – to be turned off at will when the desired level has been reached. To a unique degree, demographic conditions at any particular time are determined by what has gone before. Even if, starting tomorrow, the world's women were to have no more than two children apiece, human numbers would continue to rise substantially for another two decades. Why? Because all the world's mothers for the next twenty years have already been born, and the *numbers* entering the childbearing ages are rising year by year.

The long view requires, finally, that we acknowledge that this is a finite world: that population must, therefore, ultimately stop increasing and, in fact, that it may even need to decrease – particularly in high-consumption societies – if we are to advance toward the goal of a good life for all.

Thus, so far as optimum *size* is concerned, the dependence of human wellbeing on the interplay of so many diverse elements permits us to set only very broad limits. Recognition of the facts of ecological, resource, and social limits sets, however imprecisely, the maximum number of people who can be supported, and thereby narrows the range; but there remains, nonetheless, a considerable latitude within which this element of the demographic optimum can be located.

However, about the demographic *characteristics* of an optimum population there is much less uncertainty. If our goal is the good life for all members of society, an optimum population would have as its first characteristic a low level of mortality. It is inconceivable that death could be such a commonplace that it occasioned no sense of loss, no suffering – particularly when, in conformity to the pattern in high-mortality populations, it was so prominently visited upon infants and young children. The second characteristic of an optimum population would be a stable age and sex distribution; that is, a distribution in which there would always be the same propor-tion of the population reaching each age level: the same proportion reaching age 15, for example, in year $n+1$ as reached that age in year n; the same proportion reaching it in year $n+10$ as in year $n+9$. Such stability in age distribution obviously implies stability, as well,

125

in annual numbers of births and in relationships among age-specific patterns of mortality. Because of general economic conditions or the sheer numbers involved, it might still be difficult with such an age distribution to make adequate provision of school and other social services, and also of employment opportunities for new entrants into the labor market. But with a stable age distribution there would be no aggravation of these difficulties as a consequence of year-to-year fluctuations in the numbers succeeding to different age levels; no problems of the sort occasioned in the U.S.A., for example, by the fact that for every three children born in 1945 there were more than four children born only two years later (U.N. 1949/50, Table 11); or in Romania, the first year following the sudden clampdown on abortion, when the number of births increased 93 per cent over those of the previous year (U.N. 1969, Table 11).

The third characteristic of an optimum population would be a secular growth rate of zero. No population can increase indefinitely. There are limits: to resources, to physical space, to social space. Though these can be extended by changing the pattern of use of the environment and the pattern of behavior of individual members of society, there will be a point – even with the most judicious use of the environment and the most prudent pattern of human behavior – beyond which increases in population will inevitably result in declines in the quality of life. In fact, in consequence of the existence of limits, one could argue that a truly optimum population would have a *negative*, not a zero, growth rate; a negative rate so that human numbers – however efficient the use made of the environment – would be regularly brought into line with a steadily decreasing quantity of resources. Certainly, the current period of growth in the world – both economic and demographic – can be little more than a tiny interlude in human history.

NATURAL LIMITS

If taken advantage of through the implementation of appropriate social policies, numerical decline, or at least stability, could prove a highly useful support for efforts on behalf of both improving the quality of life and bringing human numbers, lifestyles, and consumption patterns into better alignment with ecological and physical realities. For society as a whole, the consequences of these demographic developments can hardly help but be more beneficial than detrimental.

Sweden's experience is suggestive of the possibilities. According to Birrell (1988), Sweden's attainment of population stability by the mid-1970s – with numbers distributed evenly over the youth and under-65 adult age groupings – was a major factor enabling it to channel its investment elsewhere than into such capital widening and low economically-productive activities as city building and the construction of housing. By 1984, Sweden's housing construction rate was about a third of what it had been in the late 1960s and early 1970s. On a per capita basis, it was at a rate less than half that in Australia where, partly because of higher fertility in the recent past and partly because of high immigration, the age structure has remained far from that even, near-stable pattern between successive age groupings characteristic of Sweden's. Nor, again largely because of their population stability, are the Swedes faced with those massive capital expenditures associated with population growth in major cities. For example, even excluding the costs of sub-arterial roads and certain regional facilities, and also assuming (admittedly optimistically) no increase in current price levels, Wilmoth (1988, 115–16) estimates the cost of providing the infrastructure (water, sewerage, electricity, schools, and the like) for future additions to the population of Sydney to be approximately £5,000 ($7,500) per capita. That will be investment merely to stay at the same level; it will not be investment on behalf of any improvement.

The cessation of population growth can also be expected to be at least a damper on the often ecologically deleterious practice of real estate speculation: of development by individuals on behalf of the maximization of personal financial gain irrespective of what this might entail by way of the forfeiture of ecological or amenity values. It is not impossible to control such speculation in the face of population increase, but it can be done only with effort. Birrell (1988, 103–4) contrasts Sweden with Australia, which

> has a long history, dating at least to the 1880s, of business involvement in real estate and property speculation. Continued population growth, whether by natural increase or by immigration, helps sustain this activity because it increases the scarcity value of property. . . .
>
> There is no Swedish parallel to the Australian preoccupation with property. The Swedes have not allowed investors to speculate in land or to cream off the betterment factor associated with community investment in new suburbs. Urban

land is accumulated at pre-existing rural values by local government, and allocated for building only when needed. Zoning of all urban activities is strictly enforced, limiting uncoordinated ribbon development or speculative development on the urban fringe. ...

For a variety of historical reasons, landowners were never able to gain the political power to fend off state curtailment of property rights. In Australia, once business involvement in property investment was established in the 19th century, with huge vested interests in its maintenance, it was very difficult to change. A slowing in population growth by reducing the benefits of property speculation may be an essential first step in breaking this institutional legacy.

The possibility of lower demands upon resources in consequence of a shift to older age structures is premised upon the existence of environmentally-related differences in the consumption patterns of different age groups. It warrants a closer look.

The accumulating evidence of natural limits – the damage to ozone, the insufficiency of facilities for the disposal of radioactive and toxic waste, the loss of biological diversity, atmospheric warming, and extensive soil degradation, for example – should eventually be enough to convince all but the most obstinate economic-growth ideologue of the economic and environmental truth of those two well-known social (and biological) dicta: (a) there is no such thing as a free lunch: that is, you cannot get something for nothing because benefits have their costs; and (b) you can never do just one thing because behaviors have consequences: some intended, some unintended; some desirable, some undesirable.

Growing concern over natural limits is heightened by the increasing awareness of the press upon these limits as a consequence of: (a) the continuing growth of human numbers overall, (b) the geographic redistribution of populations (especially in the form of migration to cities), and, perhaps in particular, (c) increasing levels of consumption.

Recent decades have seen substantial increases in consumption levels around the world, whether these are measured in aggregate or per capita terms. For present purposes, there are two points to raise about this increase in consumption. One concerns its human consequences, the other its future. Although it is commonly done, it is a mistake to equate these higher consumption levels with the reduction of poverty. Increased consumption can indicate declining poverty in specific instances, and, in the long run, may, of course, denote a more general reduction, as with the case of the Industrial Revolution in Europe. Yet, quite apart from the fact that measures of central tendency (and all per capita figures fall in this category) can hide gross inequalities in distribution, the more general association of high consumption with the reduction of poverty is seldom if ever above question. The English historian H. L. Beales (1958 (1928), 74) makes this point in his wry summary of England's experience of industrialization:

> [T]he broad view over a long period shows unmistakably improved standards of life for a larger population. It was a pity that so large a proportion of the workers could not measure their lives in long periods.

Another English historian, discussing the reasons for the higher military death rates among the British upper classes in World War I, refers to 'the appallingly low standards of health in many urban working-class districts' – after more than 100 years of industrial development, be it noted – and observes that 'such conditions meant that the majority of working-class men were, by the medical standards of the day, unfit to shoulder the burdens of trench warfare' (Winter 1977, 455–6). Through their poverty, paradoxically enough, they had been spared the horrors of war. Of a more recent period (presumably with particular reference to the U.S.A.), Daly observes (1977, 104):

> We have been growing for some time and we still have poverty. It should be obvious that what grows is the reinvested surplus, and the benefits of growth go to the owners of the surplus, who are not poor. Some of the growth dividends trickle down, but not many. The poor are given the sop of full employment – they are allowed to share fully in the economy's basic toil but not in its surplus – and unless we have enough growth to satisfy the dividend recipients, even the booby prize of full employment is taken away.

Working and living conditions associated with more recent experiences of industrialization suggest that the process continues much as before.

Discontent, specifically discontent with what one has, is the *sine qua non* of the growth economy. In such an economy, the

129

philosopher's goal, the contented man, is something of a bad citizen, an impediment: someone not really pulling his social oar. For it is consumption, not contentment, that forms the basis of the growth economy society. The quintessential institution of such a society – advertising – exists precisely to minimize the proportion of contented people. This it does by creating wants (Potter 1954, Ch. 8) and, in the process, by persuading people – in numbers sufficient to keep the demand side of the system operating satisfactorily – that material change is synonymous with progress, and happiness a truly purchasable commodity. The mass media's portrayals of more materially prosperous, often exotic, lifestyles but reinforce and spread more widely the discontent fostered by advertising; as does, at a more immediate level, the fact that high-consumption economic growth – particularly (although by no means exclusively) in its earlier stages, and in countries not yet industrialized – generally leads to a widening rather than a narrowing of the gap between rich and poor.

The world is not short of poor people, and their numbers increase daily. But it is not unreasonable to ask whether the lot of the poor is going to be much improved by sending them down the development path of high-consumption growth economics. Can we reasonably expect the gains to be worth the social and psychic cost? While 'development' relates to improvements in quality, 'growth' relates only to increases in magnitude. Despite the widespread assumption to the contrary – and the buttressing of that assumption by the high priority that governments accord exclusively economic matters in their national and international accounts systems (economic matters do, after all, have a concreteness and precision about them quite absent from the likes of, say, health or happiness – not to mention the appeal they have to the greedy and competitive potential in human nature) – industrial growth is not synonymous with human or social betterment. The two are not even opposite sides of the same coin. They may, for a time, coexist, but progress in the one can hinder or even prevent altogether progress in the other. Equating high-consumption economic growth with development may serve the financial or ideological interest of a few, but in a world of more than five billion people and an ever-decreasing stock of natural resources, high-consumption economic growth can offer little of enduring net value to humanity as a whole.

If the consequences of high consumption for human wellbeing can be questioned, its future can be questioned even more. Some see the spread to the rest of the world of industrialization and high-consumption lifestyles as but a matter of time and investment. It is a hopeful prediction – if you assume such development to be conducive to human wellbeing. But it is a prediction necessarily based on either an ignorance of limits or the utter disregard of them (well-summarized with illustrative quotations from some of the U.S.A.'s leading economists in Daly 1977, Ch. 5). Today's high consumption lifestyles can hardly last much longer even in those industrialized societies in which they already exist, let alone being extended to those in which they do not.

Lower consumption levels will be forced upon the populations of developed countries, whatever eventuates demographically. The existence of limits will see that it does. The market mechanism can assist in effecting this change – higher prices may, for example, induce a reduction in the consumption of fossil fuels or encourage the discovery or development of alternatives for resources that have been depleted – but it cannot do it all. Markets are not endowed with creative powers. The doctrine of 'infinite substitutability' – that the market, through ever higher prices for ever scarcer resources, will inevitably call forth satisfactory substitutes – is a matter of ecologically ignorant faith, not informed reason (for well-developed expositions of this doctrine see Barnett 1967, Australia 1973, Goeller & Weinberg 1976, Weinberg 1979).

The question is not whether changes in patterns of consumption can somehow be prevented by the shift to an older age structure, but whether an older age structure will make these changes easier and less stressful. The changes will have to be massive. Can we expect an older age structure to have a different pattern of consumption, and, if so, will the differences be of a type to assist in bringing lifestyles into closer alignment with ecological realities? In short, can we expect an older age structure to be more conducive to maintenance of a lifestyle in closer conformity to the pattern of consumption necessary to maintenance of an ecologically sustainable environment and a high level of environmental amenity?

ENVIRONMENTAL IMPLICATIONS OF CONSUMPTION PATTERNS

Four stages in the consumption process are of environmental significance: (a) production, (b) distribution, (c) consumption, and (d) disposal of waste. Environmental pollution and degradation can take place at each of these stages, and there is considerable overlap between them. The creation of waste (with attendant disposal problems) is involved at each stage; as is, also, some sort of consumption: the consumption of gas entails the consumption of roadworks, for example; the consumption of electricity, the consumption of various minerals, metals and chemical compounds.

The data

Unfortunately, so far as age patterns of consumption are concerned, the available data are few and of limited usefulness. Typical of what is available are nationwide consumer expenditure surveys in the U.S.A. and Australia (Harrison 1986; U.S. Department of Labor 1985, 1986, 1987, Australian Bureau of Statistics 1986). These have the usual limitations of survey-type data (the possibility of sampling bias, random error, inaccurate replies, unrepresentativeness of the situation at the time of interview, and the like). Moreover, the categorization by age is based on that of the 'reference person' (presumably the head of household), which means that expenditure patterns of households with a mix of ages could reduce the recorded extent of any age-related differences. In working with these data, I have tried to circumvent the possibility of bias from this quarter, so far as the Australian data are concerned, by restricting analysis to but two household groupings: (a) single-person households and (b) husband–wife two-person households; and, with respect to certain calculations I thought would be otherwise biased, have omitted household members under age 18 from calculations of per capita expenditures in the U.S.A.

More important for present purposes is the fact that these surveys refer only to the consumption of items distributed by way of the market. There is more – much more – to consumption than this, perhaps especially to consumption with environmental implications. For example, there is nothing in the surveys about sports, although the kinds of physically demanding activities engaged in most frequently by the aged (like walking, cycling, fishing, dancing, and swimming) make none of the environmental demands – extensive deforestation for the construction of runs, proliferation of roads and accommodation, snow-making and soil-compacting, for instance (see Clough 1988) – of a sport like commercial downhill skiing, the practitioners of which are almost exclusively young adults. Nor do these consumption data make any separation by gender, despite the fact that, among the elderly, women, particularly women living alone, are a larger proportion of the population at each successively older age level; and, moreover, that one can expect marked differences between men and women in a number of things – income, tastes, and health, for instance – that are likely to affect consumer behavior (see, e.g., Longino 1987).

Finally, these data pertain only to household consumption. Consumption for production purposes (the consumption, for example, of raw materials in manufacturing, or of irrigation water or oil-based artificial fertilizers in agriculture) is excluded entirely. So, also, is all 'social' consumption, which, so far as environmental impacts are concerned, means the exclusion of such important items as highway construction and military consumption; these latter involving, among other things, the diffusion of radioactive materials, the consumption of certain nonrenewable resources (especially oil) on a massive scale, the widespread introduction of toxic materials into the atmosphere, soil, and ground water, and the conversion of land, sea, and waterways to environmentally undesirable, frequently downright harmful, uses.

In short, the purposes for which these data were collected do not extend to evaluation of relative environmental impacts. Nevertheless, they do afford some insights; insights into whether a more aged population structure will be likely to increase or decrease: (a) the consumption of nonrenewable resources (b) the extent of soil degradation, (c) depletion of ozone, (d) production of CO_2 (the immediate cause of the 'greenhouse effect'), and (e) the production of waste.

Stamina and disability

On the face of it, one could expect the aged to have less impact on the environment than other sectors of the population because of: (a) lower reserves of stamina and higher rates of disability and infirmity, (b) cohort differences (in socialization and experience) of a sort likely to affect tastes and interests, (c) differences in economic condition and activity, (d) differences associated with the various stages of life.

Less stamina and higher rates of disability and infirmity could be expected to limit the aged's use of the automobile and prevent their taking part in most motorized sports: sports characterized by especially high rates of fossil fuel consumption. There is nothing in the expenditure surveys about sport, but there is something on use of the automobile.

The automobile is of singular environmental importance. A prime determinant of urban land use in industrialized countries around the world, it is a particularly important element rendering these urban areas less attractive, less habitable, less human in scale and function. It is also, of course, a major source of pollution – in many places, the single most important source: of air pollution (which is probably the most widely recognized type), but also of noise and visual pollution, and the pollution associated with the disposal of its numerous waste products, ranging from old automobile bodies to glass, batteries, upholstery, and tires. It is also a preeminent health hazard: because of its contribution to pollution and violent death, of course, but, perhaps mainly, because of the way it discourages – or in some places actually prevents (by taking up, or impeding ready access to, parklands or open spaces) – people from getting enough physical exercise.

Per capita gas consumption rises with income, but at each income level, it declines with age. The aged consistently consume less than other sectors – generally much less, especially at the highest age level (75 and over). Whether this arises from differences in taste or health it is impossible to say, but it occurs at all income levels, and there is no suggestion that the association would be altered by greater affluence among the aged. So far as this particular element of environmentally-significant behavior is concerned, the gains to be expected from an older age structure are considerable.

Vehicle purchases also rise with income – even more markedly than does the consumption of gas. But here the association with age is mixed. It is worth noting, however, that population size is *positively* associated with the pattern of outlays on vehicles at the younger ages and *negatively* so with this pattern at the older. Below age 65, the highest population concentrations are in the income categories with the highest per capital outlays for vehicles, while the lowest population concentrations are in those with the lowest per capita outlays. At the older ages (65–74 and 75 and over), it is exactly the reverse, which makes even more pronounced the differential in outlays for motor vehicles that might be expected from the changeover to an older age structure. On the American evidence (the Australian does not lend itself to such analysis), greater affluence among the aged might change this relationship at the very highest income level; but if it did, it would be only among those aged 65–74, not those in the higher age categories, and there would be no such effect at any other income/age level.

Partly as a consequence of their lesser use of automobiles, the aged are also less involved in physical violence. The result is lower rates of those injuries and disabilities associated with violence and, in particular, with motor vehicle use: costly afflictions, but more so in emotional and economic terms than in environmental.

Cohorts

The effect of cohort differences is less direct and less certain. The consequences for the environment of differences in socialization, personal experience, and experience of the kinds of social changes that are likely to affect people's tastes and interests could conceivably take several – even opposing – directions. It is not unreasonable to suppose, for example, that the current generation of old people, simply because they were reared in a less affluent time and, for that reason, presumably have less extensive material wants, could be having (irrespective of any ethical considerations or views they might have about environmental limits) a lesser per capita impact upon the environment than will – upon their attaining a comparable age – those of the environmentally-conscious youth cohort of the 1960s. Or, again because of their experience (and, again, irrespective of their views on limits), this generation of old people might have a greater personal attachment to older buildings and 'undeveloped' landscapes, and so be less inclined to tolerate the surrender of environmental amenity to the likes of coastal 'development,' road construction, and high-rise building. Whether they could translate such concern – if, indeed, they have it – into meaningful preventive action in the face of all the pressures to the contrary is, of course, another matter. At a minimum, and irrespective of their present lifestyles, older people in these industrialized countries do have at least the *memory* of having made do with less, and, for that reason, could be expected to adapt more readily to declines in rates of material consumption. This applies particularly to those born before, say, about 1930–5 in the U.S.A. and Canada, and before 1950–5 elsewhere.

But it is not beyond reason to conceive of future generations reaching old age largely, because of 'development' that will have occurred beforehand, bereft of any daily material reminders of their history; irretrievably socialized to the nonparticipatory consumption of experience (from sport to music to video sex); and with a lifetime of exposure to commercial advertising, appeals to consumerism, the encouragement to impulse buying afforded by readily-available credit, and political appeals to the narrowest, most immediate, and most self-centered of economic considerations. It is not a combination that augurs well for the development of more environmentally-appropriate behavior patterns – even if it has no connection with age structure, as such.

Moreover, as a result of environmental limits already being encountered, those now in their 20s and 30s, even some in their 40s, could well reach old age having had far fewer than will today's older people of those kinds of experiences – of a clean environment and access to open spaces, for example – likely to foster habits and attitudes appropriate to environmental realities. Because the costs – monetary, social, and personal – of continued misuse of the environment seldom come all at once, and because they may also be differently experienced by successive cohorts, it is possible that human beings, with their seemingly infinite powers of adjustment, might for a time grow used to these costs in much the way one grows used to a chronic ailment; 'inferior' or 'undesirable' conditions for one generation coming to be defined as 'normal' by the next, who would have known no other.

So far as consumer behavior is concerned, the expenditure surveys show environmentally significant cohort – and possibly life-course – differences with respect to: waste disposal, the consumption of processed foods, air travel, and package tours.

To the well-known health and safety costs entailed in the consumption of alcoholic beverages and the less well-known health costs (mainly in consequence of their high sugar content) entailed, as well, in the consumption of nonalcoholic beverages can be added the more general environmental costs entailed in the production and transport of these beverages, and the quite substantial waste disposal costs attending their use. It is worth noting, therefore, that the association between age and expenditure on nonalcoholic beverages is markedly negative (at least in Australia – the American data do not permit as detailed an analysis); as is, also, the association between age and expenditure on alcoholic beverages (after peaking among 30- to 40-year-olds).

The American surveys collected no information on either air fares or package tours, but in Australia the pattern of expenditure on air fares is one of particularly marked increases with income;

fluctuation between age groupings (with peaking in middle age), and substantially higher per capita expenditures among single-person than among husband–wife households. Where the elderly predominate in travel expenditure is in package tours: both in Australia and, according to nongovernment data (Francese 1986), in the U.S.A. The few data on the subject suggest that an improvement in their financial condition might increase travel somewhat among the younger aged and have no effect on the amount of travel among the older aged.

Social change

With regard to social change, there are a number of possibilities. We could, for example, expect the wider dispersal of kin to lead to more frequent, and more extensive, travel – either on the part of the elderly visiting their younger kin or the other way around. Some of this might, of course, be offset by the decline in the number of kin as a result of continued low fertility levels. It is also possible that a still further expansion of advertising, with its power to create wants, could produce older generations more prone to material consumption.

Dietary changes could work in the opposite direction. The surveys show the consumption of processed food (including canned, bottled, and frozen) increasing up to late middle age and declining thereafter. Modern agriculture being what it is, there may be little to choose from between fresh and processed produce so far as the effect of production practices on the environment is concerned. But with processing there is not only the charge upon the environment in the materials and practices associated with production and transport but also the potential for a further charge upon the environment in the materials and practices associated with processing itself, not to mention the waste disposal problem associated with the containers and packaging that are used. The currently lower relative per capita expenditures of the aged on processed foods must therefore be considered an environmental plus, but it remains to be seen whether this relationship will continue with successive generations more accustomed to processed (and 'fast') foods, and more responsive to ornamental packaging.

The present pattern with meat consumption is also one that increases up to late middle age and declines thereafter. But even at the oldest ages (in the U.S.A., but not in Australia, where overall

137

meat consumption levels are particularly high and there is no consistent difference by age), expenditure on meat remains higher than at ages below 45. It is possible that this is a cohort effect, a holdover of patterns of behavior developed in an earlier, more rural and less white-collar age. Succeeding generations of the elderly, reared under a different set of conditions, might consume less.

As for automobile use: while many of today's elderly – particularly the women – never learned to drive, the ability to drive is all but universal among their children and grandchildren (of either sex). While this raises the possibility of an eventual narrowing of the present gap between the aged and the rest of the society in gas consumption, it is hardly likely to result in that gap's disappearance.

Age-related economic differences – in wealth and income, property ownership, and indebtedness – can be expected to have especially important consequences for the environment. Other things being equal, the better off – with their second automobiles their vacation homes, their greater amounts of travel, and the like – can be expected to take up more land, make more demands on resources, create more waste. But, for reasons like differences in stamina and stages of the life course, this seems more likely to be the case among the younger and middle-aged affluent than among their elderly counterparts.

Stage of life can have important repercussions for consumption patterns. Among the more significant variables affected by one's stage of life are (a) the amount of discretionary time and (b) the likelihood of being in the market for certain consumer times. The aged are, for instance, less likely to be in the market for items associated with having a job (like work clothes and commuting services), or with furnishing a house or rearing children. They are, correspondingly, more likely to be in the market for items associated with having more discretionary time (like travel), as well as with having more extended periods of disability and illness. Consumption of clothing, furnishings, commuting services, travel can all have substantial effects on the environment. Whatever else can be said of it, this is much less the case with the consumption associated with disability and illness.

It is frequently claimed that old people tend more toward consolidation than expansion; more, that is, toward retention of what they have – automobile, house, clothing, friends, routines, experiences – than toward seeking out the new. To the extent such a tendency (if it exists) originates less in their stage in the life course than in lower levels of income or health, any improvement in old people's income or health will tend to increase their impact upon the environment. Judging from such data as those on motor vehicle purchases and gas consumption, however, any such increase is likely to be restricted pretty largely to the younger sector of the aged population (that is, those below 70 or 75), and, thus, to that portion of the elderly among whom the expected proportionate increase in numbers will be least.

DISCUSSION

Overall, because of their lower expenditures on motor vehicles, gas and motor oil, the aged contribute much less than other sectors to air and noise (not to mention visual) pollution, and to waste disposal problems associated with defunct vehicles and their parts. For the same reason, they have a less detrimental effect on ozone, add less to the global warming, have a less destructive impact on urban layout and amenity, add less to the sizable costs – financial and otherwise – arising out of motor vehicle crashes, and necessitate fewer outlays for those substantial socially-distributed subsidies of private motor vehicle use – like roads, traffic control, extra police and judges, higher insurance premiums, and the loss of tax-ratable property.

Compared to the rest of the population, the aged appear, also, to live a little less high on the food chain and, therefore, to have a less detrimental impact on that part of the environment involved in or affected by the production of food. It seems likely, moreover, that, because of changing dietary practices, future generations of the aged will live even less high on the food chain. Given their relatively lower expenditures on processed foods and their lower expenditures on nonalcoholic and, especially, alcoholic beverages, the aged also add considerably less to the problems and costs associated with waste disposal. Moreover, in the United States, at least, they consume substantially less electricity. (In Australia, where colder temperatures are not as widespread or severe and where so much of the space heating is done with portable electric heaters, the differences between age groupings on this score are slight.)

Although these expenditure survey data cannot show it, the aged, because of their lower stamina and higher rates of disability and infirmity, probably also make fewer demands on nonurban recreational facilities (like beaches, forests, wilderness areas, ski slopes), and so contribute less either to the degradation of these facilities or to inhibiting their use and enjoyment by others. However, with greater amounts of discretionary time, especially if this is combined with higher relative incomes, it is at least possible that the aged will, as a group, do more traveling. This would add to the rates at which nonrenewable resources are used up, and foster overcrowding and overuse of certain popular sites. But because such travel is likely to involve less extensive use of the automobile in favor of more extensive use of public conveyances – especially buses, trains, and ships – we could at least expect the additional consumption of nonrenewable resources entailed in this activity to be at a lower rate per capita.

It is possible that an older population would be less subject to fad, whether in clothing, grooming, the arts, recreation, or politics. Because young people necessarily participate less in the mainstream of the society, are less completely socialized, and have, presumably, a less extensive experience, they can be expected to experience fewer pressures to conform to adult patterns of thought and behavior. If they happen, also, to have a substantial amount of disposable wealth in their pockets – a fairly widespread circumstance in many Western countries over the past few decades – there is a further stimulus to fad in the incentive this wealth offers to exploit the greater ability of youth for economic, even political, gain. A lower proportion in the youthful age groups in conjunction with a higher proportion in the more elderly might well reduce this incentive, or at least confine it to a smaller share of the total thrust of business and politics.

CONCLUSION

Older age structures in today's low-mortality/low-fertility populations are inevitable, and numerical declines are either already under way or something that, in most of them, can be reasonably expected to be under way within a generation or two. In holding out the possibility of some lessening of the more harmful forms of human impact upon the environment, these demographic developments will present an opportunity to put our respective national houses in order in terms of both social welfare and the attainment of lifestyles more appropriate to ecological realities. If an older age structure results in less support for the economic growth ethic (as might well be the case), so much the better. But what results from these developments will necessarily be less a matter of demographic dimensions than of social policy. The threat to human wellbeing in these societies – now and in the future – originates more in nondemographic than in demographic phenomena; and, so far as demographic phenomena are involved, more in past increases in numbers than in prospective decreases. These demographic changes will present opportunities for the enhancement of human wellbeing and for significantly improving the relationship between people and the natural environment. It remains to be seen, however, whether these opportunities will actually be taken up.

REFERENCES

Australia 1973. *Economic growth: is it worth having?* Treasury Economic Paper No. 2. Canberra: Australian Government Publishing Service.

Australian Bureau of Statistics 1986. Special tabulation of data from the 1986 Household Expenditure Survey (unpublished).

Barnett, H. 1967. The myth of our vanishing resources. *Trans-Action* **4**, 6–20.

Beales, H. L. 1958 (1928). *The Industrial Revolution, 1750–1850*. New York: Kelley & Millman (London: Longmans, Green & Co., 1928).

Birrell, R. J. 1988. Can Australia flourish as a small economy? Lessons from Sweden. In L. H. Day and D. T. Rowland (eds.).

Boulding, K. 1966. The economics of the coming spaceship earth. In *Environmental quality in a growing economy*, Henry Jarrett (ed.). Baltimore, Md: Johns Hopkins University Press.

Clough, P. 1988. Alps crumbling under poison of pollution. *The Independent*, reprinted in *Canberra Times*, January 20.

Coale, A. J. 1968. Should the United States start a campaign for fewer births? *Population Index* **34**(4), 467–74.

Daly, H. E. 1977. *Steady-state economics*. San Francisco: W. H. Freeman.

Day, L. H. & D. T. Rowland (eds.) 1988. *How many more Australians?* Melbourne: Longman Cheshire.

Francese, P. 1986. Travel markets: fighting the stereotypes. In *1987 Outlook for Travel and Tourism*, Proceedings of 12th Annual Travel Outlook Forum. Washington, DC: U.S. Travel Data Center.

Goeller, H. E. & A. M. Weinberg. 1976. The age of substitutability. *Science* **191**, 683–9.

Harrison, B. 1986. Spending patterns of older persons revealed in expenditure survey. *Monthly Labor Review*, October, 15–17.

Longino, C. F., Jr 1987. The social and economic characteristics of very old men and women in the United States. Paper presented at annual scientific meeting of Gerontological Society of America, Washington, DC.

Marland, G. 1989. Fossil fuels CO_2 emissions. *C.D.I.A.C. Communications*, Carbon Dioxide Information Analysis Center, Oak Ridge National Laboratory, Winter.

Potter, D. M. 1954. *People of plenty*. Chicago: University of Chicago Press.

U.N. (United Nations) 1949/50, 1969. *Demographic yearbook*. New York.

U.S. (United States of America) Department of Labor 1985. Pre-publication data from 1985 Consumer Expenditure Survey.

U.S. Department of Labor 1986. Pre-publication data from 1986 Consumer Expenditure Survey.

U.S. Department of Labor 1987. *News*, September 24.

Weinberg, A. M. 1979. The age of substitutability. *Economic and demographic change: issues for the 1980s*, Proceedings of the conference, Helsinki 1978, Vol. 1. Liège: International Union for the Scientific Study of Population.

Wilmoth, D. 1988. The urban impact of population growth. In L. H. Day & D. T. Rowland (eds.).

Winter, J. M. 1977. Britain's 'lost generation' of the First World War. *Population Studies* **31(3)**, 449–66.

7

POLICY ALTERNATIVES: NONDEMOGRAPHIC

INTRODUCTION

Most of the concern about adjusting to an older age structure and declining numbers has focused on finances: whether there will be enough money – and enough younger people paying taxes – to meet the costs of pensions and of health and custodial services. Yet, money is not really the problem. Even if these countries fail to realize the substantial financial savings that, as already noted, are possible, none of them can be said to be in any real danger of lacking the financial wherewithal requisite to funding a reasonable and comprehensive program on behalf of its aged members. Whether such programs are funded will be a matter of social priorities and the power relations between different sectors of the society. It will not be a matter of the availability of money, as such. To make finances the prime focus of concern about adjusting to these changes is to commit the error of misplaced emphasis.

Genuine constraints on the provision of maintenance and care to the aged do exist, but they tend either to have their origin in the social and physical setting rather than the economic, or, if in the economic, to be essentially temporary in nature. Most of the present – and expected future – financial difficulty with these countries' programs for the aged arises not from increases in aging but from the relative recency of the establishment of these programs or the sudden expansion of their coverage (O.E.C.D. 1988, 69). In either case, this is the result of attempts to make up for past deficiences. In some instances, it originates in the sudden indexation of benefits to compensate for inflation: something undertaken, in the case of many of these programs, precisely at a time of unusually rapid inflation – which, while adding to the difficulty, was, after all, the main stimulus to such indexation, in the first place. The result of

143

such changes is the accrual of benefits to some individuals who have not paid for them, and the allocation to programs for the aged of monies collected from persons (usually those currently in the work force) not yet, themselves, eligible to receive the benefits for which they are paying.

While this is not an altogether satisfactory arrangment – especially, perhaps, for those paying for benefits they might never receive – it need be of no more than temporary duration. Over time, as the changes embodied in more appropriate policies are assimilated, ensuring that each successive generation of the aged very largely pays for its own maintenance and services in advance of receipt will become simply a matter of creating and administering the appropriate programs. Examples of how this can be accomplished already exist in those countries with the longer histories of welfare statism (see Ringen 1987, especially Ch. 6). Should age structures become more stable – as they will if there are no major swings in birthrates – the operation of such programs will be made just that much easier. It will be made easier, still, if there is a slowing down – or, better (were it possible), a cessation – of those types of economic 'growth' that are particularly conducive to the likes of inflation, an adverse balance of international payments, land and real estate speculation, or a widening of the gap in wealth and incomes between the very rich and everyone else. Whatever their immediate origin – whether burgeoning consumer indebtedness, the import of high-priced luxury items, foreign travel (irrespective of whether it is for business or tourism), or foreign borrowings on behalf of 'development' or corporate takeovers – the consequences of such growth, for present purposes, are much the same. These kinds of economic activity may be classified as 'growth' in national accounts systems, but they heighten uncertainty about the future and add to the difficulty of planning and saving for it (whether this saving is on an individual or a social basis). Moreover, they subject the nation's economy to a degree of control by outside forces and events over which its people can have little or no influence, and they threaten, or at least heighten uncertainty about, both the availability of suitable housing and the maintenance of neighborhood and regional amenity.

Nevertheless, these conditions are essentially temporary – even if some of their consequences are not. The more enduring constraints tend to originate in the social and physical setting – again, not in finances. The provision of personal care to the aged, for example, is more likely to be constrained by shortages of personnel to provide such care, and of time in which to provide it, than by any shortages of money, as such. These limitations, too, can be sorted out; but while finances admit of being fairly readily altered in response to changes in needs or priorities, the likes of personnel, time, or the infrastructure of cities, for example, do not. Raising wages or improving working conditions so as to attract and facilitate the efforts of additional personnel to care for the aged can help, but only very partially. Much more with constraints arising from nonfinancial than from financial limitations does ensuring adequate adjustment to the requirements of an older age structure entail changing the social and physical setting. It is such changes as these that will be addressed here. Emphasizing finances, while convenient and in conformity to the general run of current practice, simply misses the point.

CHANGING THE SOCIAL AND PHYSICAL SETTING

Introduction

There is no question that old age makes life more difficult. It could hardly be otherwise – given the disease and impairment that so often accompany it, the loss of kin and friends, the lessened control over one's life, the lower income, the lesser capacity to cope. Admittedly, those who survive to an advanced old age often seem to possess special qualities that enable them to cope with difficulties and to compensate for much of the unfavorable change they experience (A. T. Day 1991). But, significant as these special qualities may be, old people cannot in this day and age – certainly not in industrialized societies – be expected to carve out their adjustments all by themselves. To expect this of them would be neither fair nor practical. Nor would it benefit either them or the society. Old people need help – some of them more than others, of course – and this requires making some changes in the settings in which they spend the last years of their lives.

It is sometimes contended that expenditure on the aged must necessarily be at the expense of other elements in the society, particularly children (see, e.g. Preston 1984). It would be most unfortunate if it did. Apart from the question of equity, developing social programs for older people without taking into account the needs of the larger community would risk overemphasizing their plight and, in the process, also jeopardize the goal of integrating older people into the society (Gelinek 1981). It is worth noting that many of the facilities and conditions likely to improve the lot of the aged would also improve life for people at other ages, as well; perhaps most particularly for teenagers and young adults. The aged have some special interests, to be sure, but it is neither accurate nor helpful to contend that the social costs of serving these interests must outweigh the social benefits, or that these benefits will accrue more or less exclusively to old people.

Changed demographic conditions in these countries will call for some adjustments in institutions and lifestyles, partly to prevent a decline in the quality of human life and partly to raise it. But there is no need to wait upon altered demographic circumstances to pressure us into undertaking these adjustments, for most of them are already desirable on such grounds as justice, equity, human wellbeing, and stewardship on behalf of future generations. What the expected demographic conditions will do is merely emphasize the desirability of these changes, thereby offering a further incentive to embarking upon them; and, as noted, just possibly, also, make these changes in institutions and lifestyles more readily attainable.

Goals

Changing the social and physical setting to ensure adequate adjustment to the expected demographic changes entails, at its most general level, the pursuit of two types of goals: ultimate and intermediate. The ultimate goals are: (a) to enable the aged to look after themselves more effectively, (b) to enable the aged to live lives of dignity and reasonable comfort as participant, respected members of society, and (c) to develop at all levels of society those elements of personality and lifestyle, of social organization, of physical layout and functioning that are associated with being both able to cope oneself and willing and able to render assistance and comfort to others. And this needs to be done in such a way that it will result in people living in a manner that is appropriate to ecological realities and that does not compromise the ability of future generations to meet their own needs (World Commission on Environment and Development 1987, 43): living, that is, in ways that will bequeath to the generations that follow an opportunity for living the good life at least the equal of that available to themselves.

That we cannot expect these ultimate goals to be achieved either soon or in their entirety is no argument against establishing them. As R. H. Tawney observed over half a century ago (1952 (1931), 47).

> What matters to the health of society is the objective towards which its face is set, and to suggest that it is immaterial in which direction it moves, because, whatever the direction, the goal must always elude it, is not scientific, but irrational. It is like using the impossibility of absolute cleanliness as a pretext for rolling in a manure heap, or denying the importance of honesty because no one can be wholly honest.

These goals are, thus, something to be striven for; and they are also something to indicate desirable directions for change, and to serve as criteria for assessing the relative worth of alternative policies and the extent of gains and losses experienced over time.

The present discussion is therefore organized around the intermediate rather than the ultimate goals – around, that is, the means through which to approach attainment of these ultimate goals. The results of achieving any particular intermediate goal will not be specific, however. Each will serve more than one ultimate goal. What helps older people to look after themselves, for example, is also likely to develop their coping skills and facilitate their participation in society. Moreover, the consequences of achieving one or another of the various intermediate goals will be marked by considerable overlapping, with many of these goals having much the same set of consequences. We can also expect a good deal of mutual causation: changes in one direction reinforcing those in another, which, in turn, reinforce those in the first. The whole is more than a little complicated. In the pursuit of social change you can never do just one thing.

Specific intermediate goals

(a) Reduction in use of the automobile

By way of illustrating to his students how morality hinges on circumstance, a famous philosopher at New York's City College in the 1930s is said to have used the following parable: Suppose a fallen angel offered you a device that would lighten the burden of your labors and make life altogether more interesting and enjoyable. The only thing that would be required of you in return would be the annual blood sacrifice of some of your finest young men and women. You would, of course, recoil with horror, refusing to have anything to do with such a compact. Then came the automobile.

The automobile would have to be high on any list of the major determinants of lifestyles in industrialized countries – and its influence is rapidly spreading to the nonindustrialized, as well. The range of its effects extends from general employment levels to patterns of recreation and courtship, from the quality of the air we breathe and the physical layouts of our towns and cities to the future viability of the planet we live on. Its dominion over our lives is so extensive that we take it largely for granted. Yet, it is a quite recent phenomenon. Not until 1886 was one patented, and not until early this century did automobiles begin to appear in any great numbers. By 1925, however, half of the families in the U.S.A. – the most automobile-oriented country of all – owned one. Twelve years later, there was one for every 5 Americans, as well as one for every 14 Australians, 20 Frenchmen, 25 Britons, 47 Swedes, and 61 Germans. The increase since then, particularly since 1950, has been spectacular. By 1985, there was an automobile for every 1.8 Americans, 2.3 Australians, 2.7 Frenchmen, 3.2 Britons, 2.6 Swedes, and 2.7 Germans (calculated from data in U.N. 1952, Table 135, and 1985/86, Table 145). The same period has seen, as well, a marked extension of paved roads to accommodate this phenomenon: rural 'high type' surface road mileage increasing in the U.S.A., for example, from one mile for every 1103 persons in 1937 through one for every 732 persons in 1947 (U.S. Bureau of the Census 1940, Table 439, and 1950, Table 576) to one for every 323 persons in 1987 (U.S. Department of Transportation 1987, Table 5) – a six-and-a-half-fold increase in mileage alone during the period, and, what with the widening of roads and verges that took place at the same time, probably a further increase of two to three times that much, again, in land area asphalted and concretized to this purpose.

With respect to high-speed, limited access motorways, which take up more land, occasion higher noise levels, and permit higher speeds (thus, in most circumstances, entailing higher rates of fuel consumption per distance traveled), the extension in recent years has been even more rapid. Between the mid-1960s and the mid-1980s, the mileages of motorways in nonurban areas per 100 square miles of land area rose, for example, from 0.015 to 0.08 in Sweden,

0.05 to 0.47 in the United Kingdom, 0.50 to 1.25 in West Germany, 0.48 to 1.84 in the land-hungry Netherlands, and from 0.36 to a whopping 3.93 in the spacious U.S.A. (calculated from data in U.S. Department of Transportation, Federal Highway Administration 1965, 151, 164, and 1987, 122–3, U.N. 1987, Table 11).

In short, the private automobile takes up a lot of land in these countries, its waste products foul a lot of their soil and surface and ground water, and its use converts a lot of fossil fuels into atmospheric warming. So it is not a little ironic that, because of its dependence on a nonrenewable resource, the days of this voracious phenomenon are necessarily numbered. The especially high energy density of oil makes that resource particularly suitable for transport uses; but, in any long-term view of utility, these uses will have to be restricted essentially to those, like public transportation for people and rail and water transport for freight, that promise far greater efficiency and serve ends of far greater economic and social significance than does the movement of small groups of people in a multitude of individual automobiles, or relatively small loads of freight over long distances in numerous individual trucks. If substitute energy sources are eventually found (and current work with solar cells is encouraging), they can hardly be either of such a type or capable of supplying energy in such abundance as to permit continuation for long of a pattern of private vehicle usage as widespread (in terms of numbers of users and frequencies of use) or intensive (in terms of speeds and vehicle weights) as what currently exists in industrialized countries. Unfortunately, given the physical, social and ideational infrastructure centered on the automobile, reorganizing our lives to accommodate to its lesser presence – let alone its virtual disappearance – will not be an easy task.

Yet, even if the automobile did not depend on a resource that was limited, and there was, therefore, no reason from that quarter for scaling down its use, such a change in direction would be amply justified by the inherent social advantages of doing so. Some of the more important gains from the standpoint of the natural environment have already been alluded to. Some Australian researchers who undertook a detailed analysis of gas consumption in 32 major industrialized cities around the world (Kenworthy & Newman 1987, also Newman, Kenworthy & Lyons 1988) have noted that the very practices regarding land use and transport requisite to the reduction of gas consumption in cities can also 'contribute positively' to solving many of these cities' 'human, social, economic and environmental problems;' problems like: (a) 'growing frustration and pressures on personal freedom as urban living becomes dictated by longer and longer travel distances, traffic congestion and parking considerations;' (b) the 'impossibility of bicycling and walking because of the sheer distances involved and the non-viability of transit for most trips;' (c) meeting the transport needs of that large portion of the population – as much as 50 per cent in many cities – who may be described as 'transport disadvantaged,' namely, children, the elderly, the poor, and the handicapped; (d) inflation and increased balance of payments problems caused by the import of oil for transport purposes; (e) high public transportation deficits and diminishing services; (f) high urban development costs, as new roads, sewers, schools, community centers, transit services and the like are built (and duplicated) ever farther from the city center as the older inner areas of the city lose their resident populations and decay; (g) the 'loss of human vitality, intimacy and neighborliness' as areas become more functionally specialized and the mix of facilities and activities of earlier years gives way, especially in the central areas, to 'merely functional and sterile corporate centers' lacking in attractiveness and increasingly dangerous after work hours; (h) social problems of excessive privacy, isolation, and increasing crime assisted by the lack of community that an automobile induced low-density urban settlement pattern not only makes possible but actually encourages.

The urban designer, Donald Appleyard, studied the effects of automobile traffic on the residents of three blocks of San Francisco townhouses of remarkably similar appearance and with very much the same type and mix of middle- and working-class residents: one on a 'light traffic' street (about 2,000 automobiles a day), one on a 'medium traffic' street (somewhat more than 8,000 automobiles a day), and one on a 'heavy traffic' street (almost 16,000 automobiles a day). The differences found in behavior, attitude, and general quality of life were remarkable. Moving from the 'light' to the 'heavy' traffic block was to move from a broader to a narrower definition of the geographic dimensions of the neighborhood and community; from using the whole of one's house and yard, plus other areas nearby, to retreating to the confines of one's dwelling, even to abandoning parts of that dwelling, as well ('The noise has almost taken away the use of two rooms from us.'); from having several friends on the block to having few or none at all; from – and this was particularly pronounced among the older residents – having a sense of being a part of a friendly, nonthreatening environment to having a sense of loneliness and threat, instead (Appleyard, with Gerson & Lintell 1976, especially Chs. 1, 2, and 4).

To this listing could be added the encouragement the automobile gives to the development of personality traits (like aggressivenss, short-temperedness, selfishness, competitiveness, and the desire for instant gratification) which, while occasionally useful in their place and time, are more than a little likely to have anti-social consequences, and the corresponding discouragement it gives to the development of personality traits (like tolerance, cooperativeness, patience and generosity) the social consequences of which are more likely to be beneficial. We can also note the likelihood that a high rate of automobile usage, because it reduces the extent of one's contact with the physical, social, and human environment, will also militate against the development of various coping skills and traits – like self-confidence, self-discipline (perhaps particularly with respect to time – in response to the need to meet public transportation schedules – but also, given current trends toward the prohibition of smoking on public carriers, perhaps even about smoking behavior, as well) and the ability (limited through being insulated from them in the cocoon of an automobile) to encounter without fear or anxiety such elements of the social setting as novelty, strangers, human variety, and dissimilar lifestyles.

Overall, while the gains from automobile usage (the savings of time, the sense of freedom, the greater comfort, the special benefit to those in automobile-related businesses, and so on) are largely privatized, the costs (the pollution, noise, ugliness, higher insurance premiums, disability, bereavement, extra police, extra court costs, loss of ratable property, used-up resources, global warming, and the like) are essentially socialized. Everyone pays, but only a few reap the rewards.

So far as the wellbeing specifically of older people is concerned, the importance of reducing automobile usage derives mainly from the facts that:

(i) Expenditure on behalf of the automobile is such a significant part of total public expenditure that it inevitably restricts the amounts available for allocation to other areas of greater importance to the aged (not to mention other members of the society). The automobile's very considerable demands on the public purse occur directly in the form of the outlays for road construction and maintenance, and also in the outlays that, because of the automobile, are higher than they would otherwise have to be; outlays, for example, for: police, the judiciary, hospital and medical care, pensions and insurance. Indirectly, it occurs mainly through the loss of tax revenues as a result of the resumption, for the purpose of moving or storing vehicles, of land and other property, that would otherwise have been taxable.

(ii) The automobile is particularly conducive to excessively low-density urban layouts. Such layouts add greatly to the costs of establishing and maintaining such essential urban services as sewerage, electricity, and water. They also add considerably to the costs of maintaining services, like public transportation and health and community centers, that have a particular importance for the aged. Moreover, by militating against the provision of frequent public transportation and of shops and other facilities within walking distance, such layouts make it especially difficult for older people to remain active participants in the community and to take major responsibility for their own care. Those over 75 years of age (who will be some 9 per cent of the total in a stable low-mortality population with fertility at replacement level, and some 12 per cent of the total in a population with fertility 15 per cent below replacement level) are, as a group, at a particular disadvantage. Statistics on the holders of driving licenses in a prominently automobile-oriented society, the Australian state of New South Wales in 1987, are illustrative. Restricting ourselves to men (so as to avoid any cohort bias from the fact that many older Australian women never learned to drive), we find that nearly everyone has a driving license at age 25–54, and that about 90 per cent have one at age 55–69. But by age 70–74, the proportion has dropped to 80 per cent, and thereafter the decline is rapid: to 54 per cent at age 75–79, 49 per cent at age 80–84, 17 per cent at age 85–89, and 4 per cent at age 90+ (Cant 1989, Table 1). And this has to be taken as a maximum, for not everyone with a license, particularly those at the more advanced ages, will be in a condition to use it throughout the period of its validity. Moreover, as could be expected, research in such societies consistently shows that the trips made by those without ready access to an automobile are fewer in number and shorter (Baur & Okun 1983), with a particular falling off of visits to friends when those who once had ready access to an automobile no longer do (Cant 1989, 14–15).

(iii) The automobile is a major cause of environmental pollution – particularly of air, but also of soil and surface and ground-water. I have already remarked on this. So far as the aged are concerned, it is the contribution to air pollution that is especially significant, because of the particular vulnerability of older people with respiratory problems to the effects of this kind of pollution.

(iv) Use of the automobile, as already noted, would appear to have some bearing on the formation of personality structure: encouraging the development of some traits, discouraging the development of others. So far as the aged are concerned, those traits seemingly encouraged by the widespread use of the automobile tend to be ones with often anti-social consequences, while those seemingly discouraged by it tend to be ones associated with an ability to cope on one's own and a willing-ness to render help and comfort to others. The consequences, so far as older people are concerned, can be at both the individual level and the wider social level. The likelihood that older persons will be able to look after themselves, and the likelihood that they will receive the help and attention they might need or want from kin and neighbors, can both be affected. So can, at a broader level – through the influ-ence on political behavior and government administration – the likelihood that certain social programs of importance to the aged will be initiated and, once initiated, appropriately administered.

I do not want to make too much of this point. Quite obviously, there is much more than personality at work here. But the import-ance of personality should not be ignored. Moreover, it seems reasonable to suppose that anything that encourages the develop-ment of traits like aggressiveness, competitiveness, and short-temperedness, while discouraging the development of traits like generosity, tolerance, and patience, and that manages, also, to insulate people from having even visual contact with persons markedly unlike themselves with respect to, say, age or economic condition – especially when, as in the case with extensive use of the automobile, it does this on a very broad scale – has to be considered at least potentially detrimental to the wellbeing of older people. The

153

occasional drive afforded Grandmother in the ever-receding (largely because of the automobile) countryside hardly compensates for the loss of amenity and the greater difficulty she experiences getting to shops and offices or visiting friends and kin as a result of the low urban densities and the absence of public transportation that proliferation of the automobile not only makes possible but encourages. Nor, it can be argued, is the child's self-confidence and independence, tolerance, patience, or sense of curiosity fostered more by being chauffeured to music lessons, doctors' appointments, athletic events, sometimes even school, than it would be by walking or cycling to them, or if public transportation were frequent enough to permit it, going by that means, instead (see, e.g., Parr 1967, 1971).

(b) Cities built more to human scale

As already noted, achieving any of the particular immediate goals under discussion here will help achieve others, and changes in one direction can reinforce those in another. Probably nowhere is this more the case than with the two goals of reducing the use of the autombile and making cities more to human scale – making, that is, their areas of residence, work, and commerce more congruent with human needs and more suited to serving them. As a British editor recently wrote of a recent visit to Detroit:

> Detroit is synonymous with two things: cars and violent death. The Ford Motor Company and General Motors both have their headquarters there – the city was the birthplace of the American auto industry. But the city does not just make cars; it has also been made by the car. The city core is dominated by an indoor shopping mall which you can only reach by automobile. The 'Renaissance Center' is surrounded by a network of access streets which funnel cars directly into cavernous underground parking lots. Outside, the old commercial streets are largely deserted and the buildings derelict. Detroit also has the highest per capita murder rate of any city in the West. Says local resident Ralph Slovenko: 'Everything has been removed from the streets except cars and hoodlums. The more people you take off the streets, the more people become sitting ducks for crime'. As a result, adds Slovenko, a professor of law and psychiatry at Wayne State University, 'in cities like Detroit, cars are used more for protection than for transportation.' (Ellwood 1989, 4).

When it comes to the layout and functioning of towns and cities, the particular needs of the elderly would include such as the following:

(i) Ready availability of affordable, safe, and clean public transportation. In contrast with private transportation (other than walking and cycling), public transporation is not only markedly more efficient but also, as already noted, ecologically superior and arguably more supportive of those personality traits and ways of looking at things that are more appropriate to both community life and individual wellbeing.

(ii) A high degree of freedom from the physical, visual, aural, olfactory, and neurological intrusion of trucks and automobiles. This can be achieved – and already has been achieved in many cities around the world – most readily by combining the carrot of good public transportation with the several sticks of: banning motor vehicles from some areas completely and from other areas at certain times, greatly restricting the amount of parking space (limiting it pretty largely, in fact, to the infirm and those whose businesses require the use of a private vehicle), and greatly reducing vehicular speeds through redesigning roadways to the end of what has been termed 'traffic calming' (Citizens against Route Twenty 1989).

(iii) Higher-density urban layouts (but not through the means of high-rise buildings). These offer greater environmental (and economic) efficiency in the provision of public transportation, sewerage, water, electricity, and the like, and permit greater ease of mobility and readier contact with others. Although high-rise buildings might appear, at first, to meet such criteria, they are unacceptable for several reasons. For one thing, their use entails dependence upon elevators. For another, they deprive some residences and workplaces of sunlight and of views of the sky. They are also less congenial to the development of that degree of informal surveillance necessary to life in large groups; and, for some reason, vertical distances tend to be less surmountable than horizontal when it comes to the development of neighborliness and cooperation. And how high is 'high-rise?' For residential purposes, the limit should probably be no higher than about four to five storeys; no higher, that is, than would permit a parent on the topmost floor to call a child on the ground outside and be both heard and believed. For office blocks, it should probably be no higher than 10 to 11 storeys.

(iv) Readily accessible local parks, promenades, and informal meeting places that, at a minimum, are safe and clean, and, one would hope, attractive and interesting as well.

(v) Shops and eating places more to human scale. This would involve having a larger share of business in the hands of smaller-scale, more localized establishments. From the stand-point of independence and social cohesiveness, it would be preferable if these establishments, rather than being mere franchises of national or international corporations, were also completely local in their ownership and management, but that might entail their running too great a financial risk. Having a greater share of retailing in businesses of this scale could be expected to lend itself more readily to development of informal, localized security and surveillance practices (Jacobs 1961, especially Ch. 2). It might also (if only because of a lack of capital to do otherwise) foster the retention or remodeling (as against outright destruction) of older buildings of some local architectural or historical significance, and in that way heighten people's sense of community identity, continuity with the past, and predictability about the future.

Small-scale establishments could also be expected to be more readily accessible physically and perhaps (of particular significance to older – as well as younger – people), more readily accessible emotionally, as well. An establishment in which it is possble to have direct interpersonal relations between customer and shopkeeper can, in that fact alone, more readily provide a less daunting setting for someone who is a little unsure of himself, occasionally confused, possibly even a bit forgetful. It can also more readily serve some of the other sorts of functions that make for a more humane type of society. Harry Golden's (1958, 296–7) account of why he did not buy cigars by the box provides an example:

> I buy three cigars at a time, and make my purchases two or three times a day at a drugstore, a restaurant, a newsstand, or in a hotel lobby. There is no 'ritual' business. I buy them when I need them and wherever I happen to be at the moment. Thus, during the course of any week, I will have made cigar purchases in at least eight different establishments – the establishments of neighbors

in my community, in my city. This is good. Multiply that by fifty-two weeks and you'll realize how really good it is! Over the years I've made a dozen new friends, and have seen many hundreds of new people and have heard many fine new stories and anecdotes. What in the world is better than to go into a business establishment, put some money on the counter, and buy the man's merchandise! Nothing is better than that. It is good for me. It is good for him. It does something for the morale.

If attainment of this pattern of commercial establishments necessitates some broadening of our view of what is 'productive' or 'efficient,' some reduction in the number of available consumer 'choices,' even, possibly, some subsidization of commercial rents, the social gains to be had could hardly be anything but cheap at the price.

(vi) Less geographic specialization by stage-of-life and economic function. There is no denying the distinct advantages for the parents of preschool-age children to have a number of others at a similar stage of life close by – to provide playmates, assistance with childcare, advice and comfort. But this does not mean such people ought or need to be isolated from the rest of the society to quite the degree they have, on occasion, been – often, in the years since World War II, as a consequence of the rapid construction of extensive one-type housing developments on the outer fringes of our cities. Societies are heterogeneous entities. This should be recognized and made use of. It is not good for one sector of a society to be deprived of the opportunity for frequent, informal contact with the others. Certainly the aged, for their wellbeing if nothing else, need such opportunities if they are to remain participant members of the society for as long as it is physically and mentally possible, and if they are to have others nearby on whom they can rely for occasional help.

But children, in their turn, need to have contact with old people – so they will learn not to fear them; and so they will learn something of the needs and thoughts of old people, and of what to expect when their own parents, and later they themselves, reach that stage of life. Further up the age scale, the man or woman who has watched someone experiencing the death of a spouse and the period of bereavement and adjustment subsequent to it can be expected to be just that much better prepared for such an eventuality in her own life. And, it has been suggested, people of all ages – 'older persons who have regressed to childish or adolescent indifference about the effect of their acts on the people around them' along with everyone else – will be strengthened by learning to cope with diversity, by becoming 'personally aware of the milieu around their own lives' through 'experiencing the friction of differences and conflicts' within their society (Sennett 1970, Ch. 6). Just as people depend upon and learn from one another, just as they may be strengthened by having to face up to and resolve situations of interpersonal confrontation and conflict, so also can they, if conditions permit, have their lives enriched by contact with people who, in age or sex, color or income, occupation or lifestyle, are not altogether like themselves. The extent of residential specialization – the extent to which, say, single-family detached dwellings, row houses, and apartments, or rental and nonrental dwellings, are geographically mixed with or separated from one another – is a prime determinant of the availability of such opportunities.

To the desirability of decreasing residential specialization can be added the desirability, if cities are to be made more to human scale, of also decreasing functional specialization. The obviously desirable separation of dirty and noisy industrial processes from residential areas is no argument for separating commercial and governmental activities from residential, as well. Mixing residential uses with others is, in fact, one of the most effective ways of livening up the downtown areas of cities (something that three decades of constructing offstreet parking facilities has not succeeded in doing in the U.S.A., for example), and it broadens the variety of available residential types by adding a type (an apartment over or at the back of a shop or in an office building, for instance) that is likely to have a particular appeal to many elderly people, as well as to young unmarried or childless adults. Whether in the 'downtown' area or on a smaller scale within a largely residential area, this kind of mixing increases the accessibility of shops, offices, and community facilities, reduces traffic (and, therefore, all the deleterious consequences of traffic) by increasing accessibility by foot or bicycle. Moreover, because people on sidewalks are a particularly effective surveillance force it also reduces crime.

(vii) Greater diversity in available housing types and living arrangements. Not everyone is married, living with spouse, with at least one but no more than three children living at home, willing and able to drive an automobile, and with an inordinate fondness for gardening. People do get divorced or even die, children do leave home, driving and gardening do lose their appeal or become too difficult or, in the particular case of driving, too expensive. But you would never know it, to look at the acres upon acres of single-family detached houses that have sprung up on the fringes of cities and towns in a number of industrialized countries over the last several decades. These projects embody high servicing costs, frequently some environmental degradation, the loss of farmland that was once the source of fresh fruit and vegetables for the region's residents, stage-of-life segregation (as well as economic and other types of segregation), dependence upon private transportation, and residential inflexibility. The newly-widowed woman in such a setting who decides to move to a smaller more manageable dwelling must usually decide to move away from familiar surroundings, friends and neighbors, church and clubs, as well, because there is no residential alternative available. The other extreme, that of high-density apartment-dwelling (often with a design best characterized as, depending on the country, either 'Communist–Bureaucratic–Modern' or 'Capitalist–Bureaucratic–Modern') can be equally inflexible. But it need not be. It if is well-constructed, low-rise, incorporating a variety of layouts and sizes, and combined with ample public space and close proximity to facilities, it can afford quite enough flexibility to meet most people's changing needs over their lifetimes. The older areas of many European cities are a case in point, as are, also, some of the newer planned urban areas of Scandinavia. It is not that everyone has to live in one type of housing or the other; only that human needs will be better served if there is a wide choice of housing types, and if one can change from one type to another without also having to change one's neighborhood.

(c) Development of various additional social services and facilities

The social services and facilities of particular use to the aged are of two general types: those specifically geared to meeting the needs of the aged (and the infirm), and those geared to meeting the needs of persons in a wider range of ages (and conditions).

Among the first type would be such as the following: (i) home help – assistance with bathing, dressing, toileting, cleaning, laundry, and the planning and preparation of meals; (ii) household repair and maintenance services – from plumbing, carpentry, gardening, and electrical repair to window-washing and the replacement of faucet washers and overhead light bulbs; (iii) meals-on-wheels; (iv) shopping and delivery services; (v) household technical aids – like warning devices, elevators to upper floors, call buttons, and aids to telephone usage; (vi) day and recreational centers and facilities; and (vii) respite services to provide occasional relief to caregivers.

Some aged (or possibly even some infirm) persons may never need any of these; most will need – or, at least, be in a position to make use of – some of them: with a few needing them as frequently as on a regular weekly or, in the case of personal services, even daily basis, but with most needing them only on the rare occasion, as when ill or convalescing. The purpose of providing such services and facilities would be, firstly, to enhance the ability of the aged (or infirm) person to manage largely on his or her own; secondly, to assist such a person to obtain assistance when it is needed, and simultaneously to make it easier for a caregiver to monitor the condition of his or her charge; thirdly, to give caregivers some relief from the burdens and responsbilities of their task; and, fourthly, to make life pleasanter and more rewarding all around – for the aged (or infirm) person, for the caregivers, for friends, relatives, and neighbors.

Among the services and facilities from which the aged and infirm could benefit but which could directly benefit other sectors of the society, as well – especially if provided within a localized setting – are: (i) walk-in health and counselling services; (ii) inexpensive recreational facilities (for swimming, dancing, bowling, croquet, card-playing, or chess, for example); (iii) libraries (both fixed and mobile); (iv) clean, safe public spaces; (v) clean public washrooms; and (vi) public activities (like open-air band concerts, amateur theatricals, and free night-time amateur softball or soccer games). And a word could be put in, too, for the value of having (vii) a well-run neighborhood pub.

Apart from the walk-in health and counseling services, each of these would give people some exercise, get them into public areas, mix different sectors of the population together, and introduce some variety and interest, even some fun, into their lives. With people of any age, such services and facilities could be expected to help strengthen initiative and independence, even to add to a sense of community. But specifically with older people, by reducing the likelihood of their becoming entrapped in a low-expectations model that emphasized deterioration and loss of function with age, such services and facilities could help counteract the tendency to focus on one's limitations and view oneself as a burden (Phillipson 1982, 113–19), while simultaneously enriching the lives of these people and putting them in a better frame of mind for coping with life's exigencies.

So far as a walk-in health and counseling service is concerned, its major suit would be accessibility: being available when it was needed, rather than after a wait of several days or weeks; being available on an essentially informal basis without one's having to go through the process of making an appointment or getting a referral; being, ideally, accessible by foot or at least by public transportation. Members of the clergy already perform some of the functions of such a facility, but only with respect to some elements of counseling. They are neither trained nor equipped to deal with matters either more specifically related to health or, in the sphere of counseling, with finances, taxation, or the availability of public services. And, of course, a religious setting is not congenial to everyone, even for personal counseling.

Some of these services and facilities lend themselves to private provision – the neighborhood pub, for example. But most will require at least some form of public support. With those (like meals-on-wheels, shopping and delivery services, or certain of the home maintenance services) that can operate largely with volunteer labor, or which can retrieve some of their costs by charging a small fee, this support need be no more than a subsidy – for rent, equipment, maintenance, electricity, telephone, or the wages of a director, for instance. With others, however, the public will need to bear the entire cost. This will require abandoning the prejudice that exists in some quarters against having a larger public sector in the economy. A near-exclusive reliance on the profit motive and the pursuit of self-interest, on the market as an organizing principle, will not provide the variety and sufficiency of services and facilities that are needed. In many instances – perhaps most notably with automobile usage and the structure of cities, it is, in fact, very much the other way around.

Will this greater role for public provision conflict with the more qualitative aspects of how people live? Will it, for example, diminish the functions of the family, the social networks, the neighborhood, church, or private charity? Will it deprive people of arenas where they can engage in private activity and interaction? Will it lead people into a state of social helplessness, bereft of strong, human-scale institutions between the little man or woman in the street and the big government up above? A lot of theorizing has suggested it would. But the Norwegian scholar, Ringen (1987 Ch. 6), on the basis of a careful analysis of the available data on what actually happens and what people actually do – focusing particularly, although not exclusively, on Sweden, because that is the country in which the welfare system has developed furthest and in which it is expanding the most rapidly – finds no grounds for such fears. The family remains not only an important emotional community but also an important area of production (with respect to: the care of children, the elderly, the sick and handicapped; the exchange of neighborhood help and services; growing of their own food; repair services; transportation, laundry and food preparation, for example). As for nonfamilial institutions, their role has, if anything, been strengthened. 'We have here for the first time', he writes (1987, 128),

> representative quantitative information on trends in the level of individual activity in a large welfare state, covering a wide range of areas and a relatively long period of time. This information shows that the level of individual activity is high, that it has been rising during the period of welfare state expansion, that interaction among relatives and friends has become more frequent, and that the minority of the population living in passivity and isolation has become even smaller. The level of activity has increased in several areas at the same time, which means that the total volume of activity has gone up and not only that a constant volume of activity has been shifted to new areas. . . . The picture of the passive and isolated individual in the welfare state does not fit the facts. People generally have an active life style. They have become more, and not less, active.

As for the elderly, specifically, Ringen's conclusion (1987, 134) is that there are no signs in this material of a decline in family activity, of less vitality or compassion in the sensitive relationships between the elderly and younger family members, or of a decline in the quality of the family life of the elderly.

The elderly experienced significant changes in living arrangements, but these reflected not the loss of choice, but, rather, its expansion – with both generations better able to choose the arrangements corresponding to their preferences. The frequency of contact between elderly parents and their children rose, as did that between the elderly and their other relatives, while the incidence of cohabitation between elderly parents and their children went down. Families continued to provide care to their elderly members, but they were to a greater extent assisted in doing so by the expansion of social services. The increased provision of publicly provided services to the elderly in their homes Ringen sees as strengthening the family status of the aged by helping them to achieve greater independence without suffering neglect.

Quite apart from considerations like these, it must be noted that public provision is the only way some of these things can be done and, with some of them, the only way to ensure their equitable distribution to all who need them. Moreover, public provision can often provide services and facilities of a higher quality – partly because it offers at least the possibility of greater financial support and greater economies of scale, and partly because it does not require a profit and can, therefore, incorporate planning for the longer term.

A further possible gain from the provision of these facilities and services is that, by their very nature, they lend themselves to local control. Combining their provision (whether this is under public or private auspices) with user and community advisory and control bodies could assist in humanizing cities and developing coping skills and personality traits of benefit to both individuals and the society. It could create and reinforce feelings of personal competence, educate people about actual conditions within their communities and the possibilities for dealing with them, and lead to higher-quality provision of services and facilities. It could also be expected to increase social solidarity: certainly there are less expensive and arguably more socially desirable ways than war or professional sporting bodies by which a modern industrialized society can develop community feeling and a sense of identity.

(d) Less use of age as a criterion for participation in society

Most concern expressed in industrialized countries about the use of age as a criterion for social participation centers upon the labor force, particularly, now that there are controls over the exploitation of child labor, upon forced retirement at an arbitrarily fixed age. It is something of an elitist issue, however – which may explain much of the attention it receives. Where retirement is combined with an adequacy of income, it seems to hold fewer terrors, with many potential retirees actually looking forward to having more discretion over how they spend their time. The main exceptions seem to be those whose retirements involve a loss of status: the academics who have to stay in the game as Visiting Professors or as continuing participants in the lecture–workshop–conference circuit, for example, or the business executives who have to stay in it as 'consultants.'

The advantages of imposing an arbitrary retirement age are essentially three: it opens up promotion opportunities for those who are younger (opportunities that, because their timing is fixed, can be planned for), it permits the introduction of new blood (and, presumably, new ideas) without increasing staff size, and it permits the eradication of incompetence and redundancy without the unpleasantness of actually dismissing someone. Its disadvantages are that it can entail the loss of useful skills and experience, as well as productive energies, and, in addition, that it can subject workers and their families to a degree of stress and anxiety that might be largely avoided if retirement were less sudden and less age-fixed.

But there is more to age-grading than what relates to the labor force. Through introducing more flexibility into behavioral choice and increasing contacts between persons of different ages, the lesser use of age as a criterion for participation in society could be expected to enhance social cohesion and simultaneously serve individual needs. Even in industrialized societies, there is much that younger people can learn from the skills and experience of their elders – perhaps especially with respect to: interpersonal relations; adjustment to the likes of role change, loss of capacity, redundancy, death and bereavement; and (something that will take on increasing significance in the future) lifestyles more in keeping with environmental limits. And, of course, older people have much to learn from those who are younger. The gains are not all in the one direction.

Just as age grading raises barriers to the sharing of experience and culture, so should its diminution lower them. This could be expected not only to enable older persons to take a more active part in society, but, by enlarging the pool of common experience and shared culture – and, therefore, that continuous process of shared socialization that defines the members of any society – to lessen the extent of age-based group conflict, as well. In the process, it could be expected to lessen individual susceptibility to commercial and political exploitation based on the aggravation – sometimes even the creation – of conflicts of interest between age groups. Something of an extreme example of such aggravation is provided by contrasting the assiduous creation and reinforcement (at least partly for commercial purposes) of age-related differences in the propensity to follow various extremes of fashion in clothing, music, hairstyles, even drug-taking between everyone else and the swollen (and relatively affluent) 'baby boom' generation in the 1960s and early 1970s in the U.S.A. with Benet's (1974) explanation of the unusual health and longevity of old people among the Abkhasians of the Soviet republic of Georgia. Benet attributes the elderly Abkhasian pattern primarily to the cultural, social and psychological factors that structure the existence of Abkhasians, in general. Among the more important of these are: (i) a high degree of shared knowledge and common standards of ethics and morality between generations, (ii) the absence of resentment between generations (in part, according to Benet, because Abkhasian adults respect the privacy and physical autonomy of their children); (iii) the integration of older people into the life of the extended family and community as fully functioning members in terms of work, decision-making, and recreation, and the corresponding absence of age-related distinctions, like 'adolescent' or 'retired,' that limit and define conditions of existence; and (iv) the absence of sharp discontinuities between younger and older people in activities, diet, and lifestyle (Benet 1974, 103–7).

(e) Development of personality traits associated with coping and caring

Provision for human needs is a many-faceted phenomenon: a function of a complex mixture of laws, social organization, culture, geography, and the natural environment. It is also a function of personalities: of those with needs to be met, of those who play a direct role in meeting these needs, and of those who, as constituent parts of the encompassing social environment, play only an indirect role in meeting these needs. Other things being equal, because of their personality traits, some of those with needs to be met will fare well; others, badly. Similarly, among those directly servicing these needs, personality traits will produce success with some, failure with others. As for the larger society, the association of personality traits, here, will be with patterns of behavior the consequences of which relate, firstly, to the ability of people to cope on their own, and, secondly, to ensuring that support in sufficient quantity and type is available when needed, for both those in need and those who assist them.

Just what personality traits are most useful in specific circumstances is probably a matter for some conjecture; but it should not be too far amiss to reason that traits like self-confidence, self-discipline, self-respect, optimism, and an overall sense of being able to cope are going to be associated more often than their opposites with successful management of one's own affairs; or that such traits as these, together with others like generosity, cooperativeness, patience, and tolerance are going to be associated with the successful provision of care, directly, and also with general patterns of behavior, political decision-making, public administration, and so on, that are associated with maintenance of a type of social setting more appropriate to the satisfaction of human needs. I do not want to deny the occasional usefulness of traits like competitiveness, aggressiveness, the desire for instant gratification, even of short-temperedness, greed, or materialism. But their usefulness is largely confined to the individual level and to particular – and probably only temporary – circumstances. So far as the satisfaction of human needs in society as a whole is concerned, such traits will, most of the time, be more harmful than beneficial.

(f) Greater equality in the distribution of income and wealth

The more equal distribution of wealth would facilitate most of the changes enumerated above. Marked inequality of wealth merely exacerbates the difficulties to be overcome. 'What a community requires,' R. H. Tawney wrote 60 years ago (1952 (1931), 31–2),

> as the word itself suggests, is a common culture, because, without it, it is not a community at all. ... But a common culture cannot be created merely be desiring it. It must rest upon practical foundations of social organization. It is incompatible

with the existence of sharp contrasts between the economic standards and educational opportunities of different classes, for such contrasts have as their result, not a common culture, but servility or resentment, on the one hand, and patronage or arrogance, on the other. It involves, in short, a large measure of economic equality – not necessarily in the sense of an identical level of pecuniary incomes but of equality of environment, of access to education and the means of civilization, of security and independence, and of the social consideration which equality in these matters usually carries with it.

Personal wealth is a fundamentally social product. No one produces it on his or her own. Its production – and also its retention – requires society's connivance and its assistance. The significance of one's own efforts in the process is a matter of degree, of course, but the idea of someone's creating all, or even most, of his own personal wealth, the 'self-made' man (or woman) – is a simplistic myth. Tawney's still earlier observation (1920, 33–4) is as much to the point, today, as when he made it 70 years ago:

> Wealth in modern societies is distributed according to opportunity; and while opportunity depends partly upon talent and energy, it depends still more upon birth, social position, access to education and inherited wealth; in a word, upon property. For talent and energy can create opportunity. But property need only wait for it.

Societies everywhere allocate unequal shares of the social product to their respective members: sometimes in the form of personal wealth, sometimes in the form of prestige. Usually it is both, for the coin of the one tends to be readily converted into that of the other. To a degree this is defensible as necessary. To ensure the wellbeing and continuation of the society, people must be motivated to undertake the preparation and expenditure of energy requisite to meeting the society's needs; and, if they are to continue doing so, rewarded for satisfactory performance. In the nonmarket economies of Eastern Europe, for example, the necessities of life most consistently in short supply have tended to be fresh fruit and vegetables; for the production of these requires the intensive application of labor on a highly seasonal basis – something less than readily forthcoming if, as is usual in nonmarket economies, there is no opportunity to receive a higher personal reward.

But how much higher this additional reward has to be is open to question. What a man is worth is something between him and his god. What he (or she) must be paid to ensure the wellbeing of society is quite another matter. Using the pay scale of federal public servants in the U.S.A. as a guide, Herman Daly has suggested that a differential of some one to seven – the highest-paid receiving seven times the wages of the lowest-paid – ought to be enough to ensure sufficient levels of both incentive and reward for the society as a whole. In his book, *Steady-state economics* (1977, 74), he expresses the maxim another way, suggesting that no family income should be more than five times the average.

Whatever its dimensions, it is arguable that a more equal distribution of wealth serves a society better than a less equal one, perhaps especially so in industrial societies, where it can be assumed that any genuine social need for capital can be met without having to depend on rich individuals. The more equal distribution of wealth creates fewer problems and places the society in a better position to meet its needs. Where this distribution is markedly unequal, the benefits necessarily accrue to the few, the costs to the many. From the standpoint of adjustment to the demographic changes under discussion here, two types of these costs take precedence: the political (those involving the distribution of power) and the psychic (those involving emotional health).

Wealth produces power; the absence of wealth (in a society where some are markedly wealthier than others), powerlessness. Those with wealth are better able to make their voices heard and therefore, even in a democratic society, to ensure a position of dominance for their definitions of reality. The faith in further economic growth as a solution to poverty – and its complement, the concurrent rejection of greater equality of wealth as a means to this end – is a case in point. The view was succinctly put in a report by one of the advisors to the, then, American President, Richard Nixon (Moynihan 1970, 12):

There is every reason to be concerned about the costs of economic growth, and [the] need for a balanced national growth policy. ... But this is quite a different thing from proclaiming the immediate necessity to put an end to growth. ... In ... general terms, how much sense would this make for society, *given the great stabilizing role of economic growth. which makes is possible to increase the incomes of less well off groups in the population without having to decrease the incomes of others?* (italics added)

A prominent economist is equally blunt. 'Growth,' he asserts, 'is a substitute for equality of income. So long as there is growth there is hope, and that makes large income differentials tolerable ' (Wallich 1972, quoted in Daly 1977, 103). Citing this and similar views among his fellow economists, Herman Daly exclaims (1977, 103):

We are addicted to growth because we are addicted to large inequalities in income and wealth. What about the poor? Let them eat growth! Better yet, let them feed on the hope of eating growth in the future!

The point to note here is not that the idea of unlimited economic growth is silly (which it is), or that to rely on economic growth to solve problems that it itself helps create is not only irrational but also illogical (it commits the logical fallacy of *secundum quid*, namely, the omission of a significant qualification: such as that if economic growth is to be justified – as it often is – on the grounds that it is necessary to the creation of the wealth needed for meeting the costs of saving the environment, it must not be of a sort that destroys the environment it is intended to save). No, the point to remark here is that this emphasis on further growth, far from being the patter of some intellectual fringe-dwellers, emanates from the heart of academic economics and lies at the center of current political and economic decision-making. It is the view of the 'experts,' the view of those who, as one of the more prominent among them is reported to have defined it, are particularly 'capable of articulating the consensus of the people in power.' Moreover, this faith in further economic growth is all but universal. Occasionally, the qualifying adjective 'sustainable' is attached to it. But decreasing individual wealth, putting a ceiling on how far one can rise – or on how much of the world's resources one can consume – hardly rates a mention. Polite – and politically significant – discourse stops well short of such extremes.

And not only are the rich better able to keep discussion of public policy within their own terms of reference, they are also, because of their superior strength in the marketplace, better able to direct society's energies and its allocation of resources into channels from which they can most benefit. There is nothing necessarily either intentional or malicious about this; it is simply a matter of their living in a way they have learned to live under conditions admitting of material affluence – a way, in fact, that they may well have been encouraged to think of as desirable, perhaps even as socially beneficial. Nevertheless, just as it takes a lot of Africans or Indians at *their* material levels of living to consume as much of the world's resources as does the average American or German at *his* (or *hers*) (L. H. Day 1960), so also does it, in each of the societies under consideration here, take a lot of poor people to consume as much as the rich; the rich, with their vacation homes, their second automobiles, their greater amounts of travel, their higher rates of consumption overall. The rich are not all the same, of course. Nor are they the only participants in these patterns of consumption. It is simply that, to the extent they spend their wealth, theirs is the dominant position when it comes to both the consumption of resources and its concomitant, the creation of waste.

Moreover, this higher level of consumption on the part of the rich, abetted in market economies by mass advertising, also sets the pattern for much of the consumption of the rest of the society, thereby encouraging still further consumption. Among the results are total resource consumption rates higher than they might otherwise be (and thus, also, more waste to be somehow disposed of), often undesirably high levels of consumer indebtedness (because the less well off lack the resources of wealth that the rich have to fall back on), and a greater emphasis on instant or short-term gratification as against saving, the husbanding of resources, planning for the future. There is also apt to be an emphasis on the purchase of satisfaction, even on vicarious living, in preference to the acquisition of skills and knowledge – artistic, craft, social, sexual, intellectual, athletic – that could, with favorable consequences possibly enduring the better part of a lifetime, enhance one's sense of being able to cope and one's willingness and ability to take responsibility for one's own wellbeing and that of others.

This pattern of consumption grounded in marked inequality of income and wealth also nurtures privatism: the idea that to enjoy something – a beautiful view, a beach or lakefront, a work of art, a piece of equipment, the open countryside, even the opportunity to interact with children – one must own it (or them). This limits the development of those habits of sharing that, while underlying any healthy society, become increasingly necessary with the aging of its population. It also locks up for use by only a minority of its members – the extent depending on the wealth of the richest relative to that of the rest of the society – many of the resources of greatest importance to the emotional, if not also physical, health of the society as a whole; and those it does not lock up, it often destroys, so far as their use and enjoyment by anyone else is concerned – as may be seen, for example, in the boost to housing prices that has occurred in many parts of Europe in consequence of the 'vacation home' boom; or, to continue with real estate, the loss of beauty and natural amenity all over the industrialized world in consequence of the proliferation of vacation home and tourist construction (and the roads and highways servicing it) abutting lakes, rivers, snowfields, and beaches.

The issue is not one of absolute poverty, but of relative deprivation. In an industrial, high-consumption society with a markedly unequal distribution of wealth, there is both more to *feel* deprived of, because of the example of consumption set by the better-off members of the society (ably assisted by advertising), and more *actually* to be deprived of, because of what the better off are able to appropriate for their own use by way of resources, urban layout, natural amenity, and the like. A more equal distribution of wealth would preclude none of this. One could, however, expect a more equal distribution to be less supportive of these conditions and, if the top of the distribution were not, in absolute terms, unduly elevated, to lead to lower consumption rates (and, therefore, to lower waste-creation rates) overall.

Just as the markedly unequal distribution of wealth leads to higher-consumption (and therefore higher-waste) levels and the allocation of social energies and resources into channels mainly beneficial to the rich (and therefore to the detriment of those who are less than rich), so also does it entail a number of psychic costs. Especially in a society in which wealth is based, however partially, on the assumption of equal opportunity does its markedly unequal distribution lead to invidious comparisons and to discontent with what one has and frustration over not being able to do 'better.' Correspondingly, the absence of wealth under such circumstances can lead all too readily to low self-esteem and to feelings of inadequacy and failure – conditions little associated with being able to cope, or even with trying to. Marked income inequality has a way of producing incomes that are never quite high enough, especially if – as in the societies under consideration – the placement of individuals within the system is considered essentially fluid rather than fixed. In such circumstances, any satiety in consumption tends to be specific rather than general. The satisfaction of one set of wants but leads to the acquisition of another. There is always someone higher up the income scale setting the stage for discontent by consuming something to which one can easily learn to aspire. The presence of advertising makes learning the new level of aspiration just that much easier. As already noted, discontent with what one has is something of the *sine qua non* of the growth economy.

Moreover, if the pattern of income distribution is assumed to be fluid, a society with marked income differences will be a society characterized by competition. Competition has its uses, but a competitive society, by definition, is a society of few winners and many losers. Alternative opportunities in which to excel, to 'be a winner' – music, athletics, or club work, for example – can offer some compensation; but in economic matters, and also in the distribution of all those symbols of status that a competitive industrial society makes available almost exclusively through purchase (and that are also largely based on the further consumption of resources) most people will be losers most of the time.

The existence of geographic and status mobility – considered by most economic ideologues as essential to economic health at the societal level – only emphasizes this fact. As the Lynds wrote on the basis of their research in a middle-sized American city in the mid-1920s, and, later, in the mid-1930s, (Lynd & Lynd 1937, 62):

> It is characteristic of urban life, with its large jumbled populations that include many strangers, to bridge the gap between anonymity and 'belonging' by the device of overt material possessions that 'place' one. As a Middletown citizen remarked in 1925, 'People know money, and they don't know you.'

Yet, surely there is more to human motivation – in or out of the occupational sphere – than can be accounted for by the hope of invidious distinction. The satisfaction of doing something well, the approbation of one's peers, the intrinsic interest one might have in certain activities, the quality of on-the-job associations, for example. If there is, indeed, a lessening of income differences and an eventual decline in competitiveness and material aspiration as a result, we can expect a reduction in status frustration and a strengthening of such potentially useful personality traits as co-operativeness and self-respect, self-confidence, and, once again, the sense of being able to cope. Promotions and increases in income on an individual basis rather than on, say, an age or seniority basis, may well be a reward for better service or higher productivity, but they can also be socially disruptive and emotionally frustrating. They can encourage invidious comparisons, and create aspirations where formerly there were none and subordinates of those who were formerly peers. As Stouffer and his associates found in their classic study of the American soldier, morale with respect to perceived opportunities tends to be higher where promotions are few than where they are many (Stouffer *et al* 1949, 250–8). There are great emotional costs in any system, whether the work force or society itself, that condemns a high proportion of its members to the self-defined category of 'failure.'

In none of these countries would there appear to be any social necessity for the indefinite maintenance of current wealth and pay differentials. There would, in fact, appear to be considerable reason for markedly narrowing them, instead. Never mind the ethical argument for this on the grounds of fairness; the social consequences are argument enough for seeking some alteration to these systems on behalf of a better adjustment to the realities of older age structures, specifically, as well as to the needs of all human beings, generally.

CONCLUSION

Although the possibilities extend over a considerable range, and one-to-one causal relationships with specific programs are few, there is no denying that older age structures and declining numbers will require some changes in institutions and lifestyles. More will need to be done to ensure that old people are permitted a life of dignity and reasonable comfort, that society is enabled to take advantage of what older people have to offer, and that the relatively scarcer resource of children and young adults is not wasted. There will be a need for fewer age-based restrictions, for greater flexibility in retirement and in work times and practices, and possibly for more equal wage structures so as to attract workers to certain less desirable, but necessary, jobs. There will be a need, also, for better provision of certain social services and – contrary to some present trends – for the extension, rather than contraction, of social provision for many needs formerly met through the efforts of kin; for, not only has the world changed economically, but, in consequence of low fertility, family networks will, on average, be less extensive and, for a possibly larger proportion of the population, nonexistent altogether.

Overall, there will be a need for planning to enable those in the enlarged age sector to take care of most of their noneconomic needs without undue reliance on others. Fortunately, many of the services and facilities this would require – good public transportation, walk-in health maintenance and counseling services, public spaces where one can be by oneself or meet people on an informal basis, inexpensive rental housing, for example, – are also of particular use to others in the society, perhaps especially to children and young adults. It is not just old people who stand to benefit from their development and extension.

In its 1978 Annual Report, the United Kingdom's Supplementary Benefit Commission eloquently defined a minimum set of conditions with respect to poverty (Supplementary Benefit Commission 1978, 2, quoted in Ringen 1987, 148–9): 'Poverty, in urban, industrial countries like Britain,' they wrote,

> is a standard of living so low that it excludes and isolates people from the rest of the community. To keep out of poverty, they must have an income which enables them to participate in the life of the community. They must be able, for example, to keep themselves reasonably fed, and well enough dressed to maintain their self-respect and to attend interviews for jobs with confidence. Their homes must be reasonably warm; their children should not feel shamed by the quality of their clothing; the family must be able to visit relatives, and give them something on their birthdays and at Christmas time; they must be able to read newspapers, and retain their television set and their membership of trade unions and churches, and they must be able to live in a way which ensures, so far as possible, that public officials, doctors, teachers, landlords and others treat them with the courtesy due to every member of the community.

It is much the same with the needs of old people (or anyone else, for that matter, regardless of age or economic condition). They should be adequately nourished, comfortably housed, able to be self-respecting, treated with courtesy, and, rather than excluded or isolated, enabled to maintain contact with their kin and to participate in the society.

Although there may be some differences as to weighting in specific instances, in the essentials old people would be best served by much the same sorts of conditions that would benefit the rest of society. In the most general terms, what is needed is a *livable society*; a society that, as a minimum: (a) is built, in terms of its physical layout, and structured, in terms of interpersonal relations, to human scale; (b) gives priority in physical movement to pedestrians and cyclists rather than motorized vehicles and, so far as the latter are concerned, to public rather than private carriers; (c) enjoys a high degree of behavioral predictability; (d) is safe; (e) is free of air, water, and soil pollution; and (f) has in operation a high-quality safety net with respect to housing, health and custodial care, rehabilitation and counseling services.

As noted above, these are to be seen more as goals to be striven for, and as criteria for assessing the worth of social policy, than as something to be attained in any imminent future. This is no denial, however, of either their inherent worth or their relevance to the needs that can be expected to gain particular prominence with the transition to older age structures.

Any attempt to achieve such goals must necessarily clear a formidable array of obstacles, ranging from the physical (such as the current layouts of towns and cities) through the institutional (such as the economic system and the system of social stratification) to the ideational (such as people's assumptions and priorities about what is acceptable or unacceptable, good or bad, desirable or undesirable). Yet, countervailing forces do exist, if not to a degree that justifies much optimism, at least to one that justifies cautious, if limited, hope. These relate to: (a) the turnover of physical stocks, which offers the possibility of either not replacing at all (as with some roads or buildings that occupy sites that could be put to less socially disadvantageous uses than at present) or of replacing with structures less socially disadvantageous; (b) the turnover of human generations, each new generation offering at least the possibility that through education and experience it can become an improvement upon its predecessor; (c) the possibility of changing people's priorities and behaviors through, say, better informing them about social and environmental conditions, and affording them greater opportunity for – and reinforcing experience of – behaving in ways more concordant with social and environmental requisites; and (d) the possibilities offered by political structures for effecting wide-ranging improvements in, among others, social services, environmental quality, and the availability of public transportation (see L. H. Day 1982).

REFERENCES

Appleyard, D., M. S. Gerson & M. Lintell 1976. *Liveable urban streets: managing auto traffic in neighborhoods*. Washington, DC: U.S. Department of Transportation, Federal Highway Administration.

Baur, P. A. & M. A. Okun 1983. Stability of life satisfaction in late life. *The Gerontologist* 23(3), 261–5.

Benet, S. 1974. *The long-living people of the Caucasus*. Rinehart & Winston.

Cant, R. 1989. Cars and the social networks of the elderly: the creation of disadvantage. *Australian Journal of Ageing* 8(3), 11–16.

Citizens against Route Twenty 1989. *The Solution to Route 20 and a new vision for Brisbane*. Ashgrove, Queensland: C.A.R.T.

Daly, H. E. 1977. *Steady-state economics*. San Francisco: W. H. Freeman.

Day, A. T. 1991. *Remarkable survivors: insights about successful aging among women*. Washington, DC: Urban Institute Press.

Day, L. H. 1960. The American fertility cult. *Columbia University Forum* 3(3), 4–9.

Day, L. H. 1982. Changing resource use patterns: the social possibilities. In *Quarry Australia? – social and environmental perspectives on managing the nation's resources*, R. Birrell, D. Hill & J. Stanley (eds.). Melbourne: Oxford University Press.

Ellwood, W. 1989. Car chaos. *The New Internationalist*, No. 195, May, 4–6.

Gelinek, I. 1981. The World Assembly on the Elderly from the point of view of the International Council on Social Welfare. In *The U.N. Assembly on the elderly, the aging as a resource; the aging as a concern*. Proceedings of two meetings organized by the International Federation on Aging, May 27–8, 1980, Vienna. Washington: International Federation on Aging, 12–15.

Golden, H. 1958. *Only in America*. New York: World Publishing Co.

Jacobs, J. 1961. *The death and life of great American cities*. New York: Random House.

Kenworthy, J. R. & P. W. G. Newman 1987. Learning from the best and worst: transportation and land use lessons from thirty-two international cities with implications for gasoline use and emissions. Transport Research Paper 7/87, School of Environmental and Life Sciences, Murdoch University, Perth, Western Australia.

Lynd, R. S. & H. M. Lynd 1937. *Middletown in transition*. New York: Harcourt Brace.

Moynihan, D. P. 1970. Counsellor's statement. In National Goals Research Staff, *Toward balanced growth: quantity and quality*, Washington: U.S. Government Printing Office.

Newman, P. W. G., J. R. Kenworthy & T. J. Lyons 1988. Does free-flowing traffic save energy and lower emissions in cities? *Search* (Australian and New Zealand Academy for the Advancement of Science) 19(5/6), 267–72.

O.E.C.D. (Organisation for Economic Cooperation and Development) 1988. *Ageing populations: the social policy implications*. Paris.

Parr, A. E. 1967. The child in the city: urbanity and the urban scene. *Landscape* 16(3), 3–5.

Parr, A. E. 1971. To make the city a child's milieu. *New York Times*, July 4.

Phillipson, C. 1982. *Capitalism and the construction of old age*. London: Macmillan.

Preston, S. H. 1984. Children and the elderly: divergent paths for America's dependents. *Demography* 21(4), 435–57.

Ringen, S. 1987. *The possibility of politics: a study in the political economy of the welfare state*. Oxford: Oxford University Press.

Sennett, R. 1970. *The uses of disorder*. New York: Random House.

Stouffer, S. A. *et al.* 1949. *The American soldier*, Vol. 1. Princeton, N.J: Princeton University Press.

Tawney, R. H. 1920. *The acquisitive society*. New York: Harcourt Brace.

Tawney, R. H. 1952 (1931). *Equality*, 4th edn. London: George Allen & Unwin.

U.K. (United Kingdom) Supplementary Benefit Commission 1978. *Report*.

U.N. (United Nations) 1952, 1985/6. *Statistical yearbook*. New York.

U.N. 1987. *Annual bulletin of transport statistics for Europe, 1985*. New York.

U.S. (United States of America) Department of Transportation, Federal Highway Administration 1965, 1987. *Highway statistics*. Washington, DC: U.S. Government Printing Office.

Wallich, H. C. 1972. Zero growth. *Newsweek*, January 24, 2.

World Commission on Environment and Development 1987. *Our common future* (The Brundtland Report). Oxford: Oxford University Press.

8

CONCLUSION

All in all, a society's conditions of life are unlikely to be much affected by either the number of the elderly or their share of the total population. Far greater importance can be expected to attach to: (a) the physical and mental condition of the elderly themselves and (b) the political, economic, geographic/environmental, and social conditions in the society of which they are a part. Such elements are, of course, interactive. Old age will be burdensome – both to the aged and to the society – to the extent that housing, medical attention, income distribution, and the availability of public transportation, for example, are not of a sort to enable the great majority of old people to remain active and useful, assured of status and livelihood, and enjoying a reasonable degree of predictability about the future. Old age will be burdensome if the elderly are not enabled – as they so often can be – to look after most of their needs themselves; if those who provide them with care – whether kin, friends, or professionals – are not themselves provided with the facilities and assistance necessary to that task; if cities and towns, parks and recreational areas are not laid out in ways appropriate to pedestrian, rather than motorized, use.

In short, old age will be a problem to the extent that: (a) societies do not operate at a human scale, (b) social policy focuses more on increasing the size of the economic pie than on equitably distributing it, and (c) both natural and human-made environments are maintained at a less than high-quality level. Old age will be a problem to the extent, that is, that social and environmental conditions fail to meet the needs of all age groups (and of future generations no less than present) – with some extra attention to the special needs that arise at particular stages of the life course, old age included.

Older age structures in these low-mortality/low-fertility populations are inevitable, and numerical declines either already under way or something that can be expected to be under way in most of them within a generation or two. But what results from these developments will be a matter less of demographic dimensions than of social policy. The expected demographic changes in these countries could greatly assist efforts to put their respective national houses in order from the standpoint of both social welfare and the attainment of lifestyles more appropriate to physical and ecological realities. But whether efforts in these directions will be forthcoming and this possibility of assistance taken up remains to be seen. There is no assurance that the societies concerned will respond appropriately to the challenge. Their responses could be rudimentary, too narrowly focused, based on too short a time frame.

But this, for the present, is beside the point. What is to the point is: firstly, that a lot can happen before these populations experience numerical declines to any arguably serious extent; and, secondly, that to wring one's hands over present demographic trends in these countries is not only to commit the error of misplaced emphasis but to risk the enactment of irrelevant or undesirable social policies, as well. The concern over these trends appears to be based less on an awareness of their likely consequences than on the fact of moving on to unfamiliar demographic ground before social attitudes and institutions are quite ready for it.

These countries still have more to lose socially and environmentally from past *increases* in population size, and *unevenness in age structures* in consequence of past fluctuations in birthrates, than they do from any near-zero population growth rates or the transition to persistently older age structures. We must beware of determinisms – demographic no less than any other. Human wellbeing in these countries is threatened more by nondemographic than demographic phenomena; and, so far as demographic phenomena are concerned, more by past increases in numbers than by either prospective numerical declines or the trend to older age structures. To the extent that demographic conditions are, themselves, determinants of social conditions, the older age structures and smaller numbers expected in these countries are far more likely to be desirable than undesirable. This is so from the perspective of the individual and the society; it is so, also, from the perspective of the world.

INDEX